the soil
will save us

HOW SCIENTISTS, FARMERS, AND
FOODIES ARE HEALING THE SOIL
TO SAVE THE PLANET

the soil
will save us

KRISTIN OHLSON

RODALE.

For Holland, Sylvie, and all the inheritors:
Don't be meek!

Rodale books may be purchased for business or promotional use or for special sales.
For information, please write to:
Special Markets Department, Rodale Inc., 733 Third Avenue, New York, NY 10017

Printed in the United States of America
Rodale Inc. makes every effort to use acid-free ♾, recycled paper ♻.

Book design by Amy C. King

Library of Congress Cataloging-in-Publication Data is on file with the publisher.
ISBN 978–1–60961–554–3 hardcover

Distributed to the trade by Macmillan
2   4   6   8   10   9   7   5   3   1   hardcover

We inspire and enable people to improve their lives and the world around them.
rodalebooks.com

CONTENTS

*"Even the broken letters
of the heart
spell earth."*

—DANIEL THOMPSON

INTRODUCTION

I'm working in my Cleveland backyard, but idly, dawdling in the sights and smells of fall. A starry canopy of yellow leaves sets off a vivid blue sky. The already fallen, the now-brown leaves, steep in damp, shadowed corners of the yard, and their lapsang-souchong fragrance blows my way. Only the annual chorus of leaf removal mars this otherwise peaceful day. In some yards, blowers roar and raise poofs of dust. In others, people scritch and scratch with their rakes. I'm of the scritch-and-scratch persuasion, wielding an aged plastic rake with many broken tines, like bitten-down fingernails on a large green hand.

Normal enough, except that I'm not raking my leaves onto a tarp to dump into a recycling bag or to drag to the curb, as my neighbors are. I'm raking the leaves on my driveway back to my lawn, hoisting big piles of them onto the muddy turf and then smoothing them with the rake so that they cover the sparse grass with a thin, multicolored coat of autumn. I can imagine my first ex-husband howling, "You're going to kill the grass!" No, I'd tell him; I'm trying to *save* the grass.

I don't care much about having a good lawn—women

don't, in my experience, and men do. I remember my father in his eighties, surveying the sweep of emerald in back of my parents' house and sighing, "I just want a perfect lawn before I die." My siblings and I found this both hysterical and poignant. Hadn't he always had a perfect lawn? And since it probably wouldn't get much better, was he doomed to die disappointed?

I've always viewed my lawn as just the blank space between my flower beds and the oval of soil where I grow vegetables—a spot that benefitted from all-day sun when we bought the house decades ago, but is now shadowed most of the day by oaks and maples and the occasional doomed elm. The grass wasn't especially lush when we moved in, and our tenure further traumatized it. We had a drainage problem after we put in a new concrete driveway and garage: every time it rained, a foot of water backed into the garage. So the yard was torn up as our contractor dug one French drain after another to fix the problem (they didn't), and then another contractor finally tore up everything, driveway included, to route all that storm water out to the sewer.

So the backyard was crushed by heavy equipment off and on for 2 years. If the moon were made of gouged and gridded mud, my view out the kitchen window looked like a moonscape. For Easter during one of those years, I bought my kids a packet of "crazy gourd" seeds to plant. Vines soon covered the entire yard, blurring the ugliness with their curvy, slightly furred leaves and relieving me of the heartbreak of planning a garden, only to have to summon the bulldozers in again. (Someone ought to try making ethanol from those crazy gourds; I've never seen such miles of vegetation grow from a tiny handful of seeds.)

We finally planted the lawn and flower beds. The flowers were great—a rollicking seasonal carnival of colors and shapes—but the lawn really never had much of a chance. Too much compaction of the soil by the bulldozers. Too many seasons of Slip 'N Slide. Too much running and jumping, and too much use of stilts and pogo sticks and Big Wheels. Too many basketball games that veered off course. Too much peeing by one dog and digging by another. When the first marriage ended and the second began, a wedding with 100 guests tromped that weary turf for 3 celebratory hours. Then two more dogs—the young black ones that crash through my piles of leaves now—chased each other back and forth across the grass, their nails flinging tufts of grass in their wake until nothing remained but gouge. Also the dearth of water. Even though Cleveland is a high-precipitation city, there are times when you have to water, and honestly, I've always been stingy with the grass.

So I find myself now with a lawn that is mostly just exposed soil. So hard when it's hot that you could break a plate on it; so muddy when it rains that I'd rather walk the dogs in a downpour than turn them loose in the yard. The dogs are my only companions now, my sole beloveds-in-residence, since the second marriage has ended and the kids have moved out. Finally, I too pine for a good lawn, if only to keep the dogs from getting muddy.

Early in the fall of 2011, I scoured the newspaper for instructions on pre-winter lawn care, wrinkling my nose at the ads for lawn chemicals—I'm categorically opposed to them, and look how they broke my poor father's heart anyway. An article written by someone from the Cleveland Botanical Garden recommended aeration and compost and reseeding, but I only have enough compost for one small corner of the yard. I spread the compost there, jab a pitchfork every few inches to aerate, then cover the rest of the yard with leaves. I might get a better lawn from this next spring. And I might do my tiny, infinitesimal part to heal our climate and nurse a number of other ills that have their secret roots in the soil.

I first heard about the connection between soil and climate 3 years ago. I had written a feature article* back in 2005 for *Gourmet* magazine about a local restaurateur named Parker Bosley: He not only had earned two of their top-chef shout-outs, but he was also a pioneer locavore, who began searching out local ingredients for his restaurant back in the late 1980s. He was raised on an Ohio dairy farm, became a teacher, spent time in France, and became smitten with the menus there that evolved with the seasons, according to what was fresh and at its best.

He began to replicate that when he started his restaurant in Cleveland. He stopped at farm stands outside the

---

*Sadly killed before it made it to print

city, tasted their peaches, and asked, "If I come back next week, can you sell me a few bushels?" He ventured out into the countryside of his youth, knocked on the farmers' doors, and told them he wanted to purchase their pork or eggs or chickens, no middleman. He urged farmers to try heritage breeds, to let their pigs live in the woods and eat acorns and apples, to give their chicks bowls of soured milk so they could pick at the curds. Soon, he had a pipeline of locally produced foods for his restaurant—by the time I interviewed him, 97 percent of what he served was locally sourced—and to the farmers' markets that sprang up around the city. He influenced what was happening on small farms near Cleveland, helping many of them survive and even expand. He always knew about the family that was starting to make sheep's milk cheese near Toledo, and the French butchering techniques that would yield better cuts of meat, and even about the young guy from Tennessee who was in Italy, learning the art of making *salumi*.

Every time I needed a new topic for an article about food, I'd call or e-mail Parker and he'd let me know what was new. And one day, he said, "Carbon farming. That's the new thing."

He explained that there was a new movement among small farmers. They were changing their practices with the soil in both large and subtle ways, knowing it was the foundation of all agricultural life, whether they raised chickens or corn, pigs or spinach, beef or peaches. Sometimes they called themselves soil farmers. Sometimes they called themselves microbe farmers, aware of the billions of tiny creatures that they couldn't see but that scientists told them were at work in the soil. Sometimes they called themselves carbon farmers, knowing that it was carbon that was making their soils richer, moister, and darker. Some had been following the work of scientists who said that this kind of farming accelerated the removal of carbon dioxide from the atmosphere via photosynthesis and could slow and maybe even turn back global warming. The ones who believed in global warming took pride in this. The ones who didn't—and there are many in agriculture who still don't—were nonetheless thrilled to see their lands, crops, and animals thrive in ways they'd never imagined.

I followed the activities of some of these scientists, carbon farmers, and carbon ranchers for a few years before I started this book. I went to their conferences, read their blogs and scientific papers, inspected their fields, ate their produce, yammered about them to my friends, and wrote about them. Right away, I was stunned by what I learned about life in the soil—that when we stand on the surface of the earth, we're atop a vast underground kingdom of microorganisms without which life as we know it wouldn't exist. Trillions of microorganisms, even in my own smallish backyard, like a great dark sea swarming with tiny creatures—it almost makes me feel a little seasick standing there, knowing how much business is being conducted right under my feet.

One of the principles I took away was that bare land starves the microbes below the soil. These microbes need living or dead plants to get their foods—their sugars, carbohydrates, and proteins. They'd prefer a thick and diverse clump of growing plants—roots in the ground!—but dried-up biomass will keep them going until the juicy stuff returns. Thus the idea for raking leaves over the top of my naked lawn. Maybe if I give my microbes dead leaves to chew over the winter and into next spring, they'll thrive and begin to aerate the lawn themselves as they burrow through the soil and spread their dark constellations underground. Maybe the lawn will then be more receptive to seed and water next spring.

Or maybe not—maybe it will take more complicated efforts to restore my lawn, just as it will take more complicated efforts to restore carbon to the soils of the earth. But it was my own little experiment, nonetheless. I just hoped one of my neighbors didn't see the piles of leaves in my yard and send one of his sons over to rake it up, just as other neighbors have turned up now and then to shovel the snow from my driveway. It wouldn't make sense, all those leaves on the lawn, unless you knew something about the incredible life in the soil.

CHAPTER 1

WHERE DID ALL THE CARBON GO?

Plot 87 of the Waterman Agricultural and Natural Resources Laboratory at Ohio State University used to be part of a farm. Pioneers chopped it out of the dense central Ohio forest in the 1800s. They raised corn, wheat, and oats for the horses; rye for making whiskey; flax, which they mixed with wool to make linsey-woolsey clothes for the men; apples, perhaps even a variety carried to Ohio by John "Johnny Appleseed" Chapman; and probably a dozen more crops. It was enough to feed themselves amply from the rich soil below, have some to share with neighbors, even sell to the occasional stranger. They picked arrowheads and stone beads from the soil, wondering about the ancient native peoples who had also flourished in the valley's great fecundity.

The Watermans were the last to plow this land, and when the city and suburbs of Columbus closed in on their farm they decided to give it up to the university. Now only a few dozen acres remain, encircled by swaths of forest, which are in turn surrounded by the new urban canyons.

Two unlike groups of moving objects, the nearby traffic and the hundreds of birds that swoop and survey the fields, make a cacophonous backdrop.

These acres have been preserved from the bulldozers so that a handful of soil scientists can pursue experiments. On a cold, wet morning of a spring that had been cold and wet for weeks, Rattan Lal drove me there to see what, if anything, was growing. He was afraid that it would be too early for their test crops to have popped the sodden crust of soil.

But as he stepped from the university truck and navigated the puddles on the dirt road, he smiled and pointed. "There is something growing!" In Plot 87, rows of new corn—like wavering lines of tiny green feathers—stretched toward the blocky city skyline.

Lal is the director of the university's Carbon Management and Sequestration Center (C-MASC), which attracts researchers from around the world. The experiments surrounding us were those of his students. He has been at this work for 50 years. A tall, graceful man with white hair and large gray-rimmed glasses, he doesn't dig in the dirt anymore.

Even so, whoever planted this particular field relied on two of the methods for enriching the soil and preventing erosion that Lal came to appreciate early in his career and has preached the use of for years. No-till agriculture, for one. I grew up in an agricultural valley in California and love the rhythm of plowing, the graceful lines incised on the land, the opening up of all that bountiful and mysterious earth. I especially love it in Ohio, which was still my home when I visited Lal in 2012, where you can drive down back roads and see Amish farmers plowing with their teams of giant, furry-hoofed horses. But plowing actually damages the soil structure and exposes soil carbon—the crumbly blackness that generations of farmers have recognized as a feature of the best, richest soil—to the air, where it combines with oxygen and floats away as carbon dioxide. So this field was planted using a machine that punched slits in the soil through the roots and debris of last year's crop and dropped in seeds. It was a field without furrows.

The second difference: While the thousands of acres I passed on my way down from Cleveland rolled along the sides of the road like bolts of neat brown corduroy, Plot 87 was littered with bits of dry leaves and cornstalks. Instead of burning the residue from last year's crop or letting someone haul it away for their pigs or sending it off to an ethanol plant that previous fall, the residue had been chopped up and spread over the land. Left there on the ground, the residue reduced erosion. It kept the soil temperature cooler during the summer. It provided food for worms and other creatures that aerate and enrich the soil, and thus help make it more porous and absorbent.

Lal leaned down and tapped a long, slender finger on a layer of half-rotted corn stalks, the swirling *om* tattoo on the top of his hand a faint blue in the weak sun. He pushed the crop residue aside to examine the soil. "See how the ground underneath has no cracks?" he said. "When it's covered, the residue protects it from the hot sun and it doesn't dry out so much. And here"—he poked his finger into a loose mound of soil—"here, you see where organisms have been eating the residue. This is their waste product, which makes the soil richer and loosens the soil. An earthworm can drag a leaf down more than three feet into the soil."

His gaze switched from the micro to the macro. "When you look across the field, you see that there is no water on the surface," he said. "This soil absorbs moisture. The water doesn't run off or make puddles."

I recalled seeing the pools of water from last night's rain on the fields along the highway on my drive south. This field was like a huge brown sponge, with all that water lying beneath our gaze, suspended in the land's pores.

It was what I couldn't see that had drawn me to meet with Lal, and that was the carbon content of the soil. Lal became a soil scientist mindful of the world's poorest farmers—he grew up among them, first on the Pakistan side of Punjab and then, after partition, on the Indian side—thinking only of helping them raise better crops and save their soil from washing away. Along the way, he realized that when the soil lost its life-giving carbon through plowing and poor land management, its disappearance into the air wasn't just a blow to the millions of poor farmers or an inconvenience to the thousands of corporate farmers in the developed world, it was adding substantially to the threat overhanging the entire planet: the load of greenhouse gases in the atmosphere. In fact, up until the 1950s most of the excess carbon dioxide in the air resulted from the ways humans used their land and forests.

So even though he doesn't dig in the dirt anymore, Lal stays busy. He spends much of his time traveling to conferences around the world, talking about the connection between soil carbon and global warming. The message: We need to do everything we can to stop losing historic soil carbon, and we also need to do everything we can to build and retain more carbon in the soil. He also takes this message to the National Climate Assessment and Development Advisory Committee (NCADAC), where he is the only soil scientist. Other soil scientists know about the connection between soil carbon and global warming, but they have not been aggressive about taking this message outside their field or explaining the importance of their work.

"Even I am bad about this," said Lal, who was the president of the Soil Science Society of America (SSSA) in the early 1990s. He grinned. "The other day I was explaining something to a group and I said, 'It isn't rocket science.' I thought about this later. I should have said instead that 'It isn't *soil* science'!"

Even though Plot 87 is a far cry from his family's 2-acre farm in India, it brings back memories. How his father plowed the fields with a team of bullocks, and then he and his father leveled the fields by sitting on a board, which the bullocks dragged over the ground. How they separated out the wheat seeds by drying the wheat in the sun and then nudging the bullocks to drag it over the ground. How they cleaned the seeds by throwing them up in the air and letting the wind whisk away the dirt and chaff.

They lived in a village of mud-brick houses, with no electricity or even roads. All the men wore beards because no one owned razors and there wasn't a nearby barber. One came to town monthly and trimmed hair and beards in exchange for a share of the wheat and rice crops. Every family had a cow for milk, which they converted to yogurt, a staple of their diet. When a cow died—all good Hindus, they'd never think of killing one for its meat—the butcher arrived to convert the hide into shoes. "They were so stiff and painful for the first few days!" Lal said. "We always got blisters. Sometimes we just carried the shoes around in our hands."

They had enough to eat on the farm, just barely, but starvation was rampant in the cities. Farmers just didn't produce enough to feed the population. Looking back with a scientist's eye, Lal sees that low, dwindling productivity was inherent in their centuries-old methods. It was not only the plowing that was a problem. The greater concern was that the villagers took from the land and never gave anything back—something he now calls "extractive farming." They gathered crop residue from the fields and took it away to burn in their stoves or to sell at the market. They gathered the bullocks' dung, dried it, and burned that, too. Leaving both in the fields would have enriched the soil but further impoverished the farmers. "The residue and the dung were precious, and we carried them away," Lal says. "No one leaves them on the ground there, even now. They're still precious to a poor farmer."

When he was 7 years old, a peddler bicycled through the village offering tattoos for two pennies. Lal doesn't remember where he got the two pennies, but he had the peddler tattoo his hand with the symbol for *om*, a Sanskrit word meaning the sound of creation. As he looked at the small black welt, he felt that this wasn't enough for the mighty sum of two pennies. The peddler obliged him by tattooing his initials—*RL* for "Rattan Lal"—on his upper arm. The boy couldn't read English, so the letters meant nothing to him.

But he knew om. His grandfather was a Hindu priest, and the boy knew 40 yoga poses and how to recite the mantras that went with them. The family intended him to follow his grandfather into the priesthood, but the youthful Lal knew more than his mantras—he was also good at math. Finally, the family decided to send him to university in Delhi. He was wandering the halls one day and struck up a conversation with a young man crouched next to a doorway. He was a peon—it was an actual paying job, to run someone else's lowly errands—and Lal asked him if he could help him find a job as a peon, too. Just then, a jeep emblazoned with the name OHIO STATE UNIVERSITY pulled up outside the building. "Americans are here?" Lal asked.

The peon nodded. "And they give scholarships."

A few years later, Lal arrived at Ohio State on scholarship

with an $8 traveling stipend from the Indian government—
a sum he found lavish until he discovered that a night's
housing cost more than that. He found himself a place to
stay by following the sound of applause down a university
corridor and peering in to see a roomful of international students. Some were wearing the turban of his native Punjab,
and they invited him to stay with them until he got on his
feet. After that, there were more scholarships and awards
and mentors, enough to assure his brilliant future. "I am
very lucky," Lal says. "The stars aligned for me."

Years passed before he tripped over the first big hurdle
in his career. After he finished his PhD in soil science, he
was hired by the International Institute of Tropical Agriculture in Nigeria, 1 of 15 research centers funded by the
Rockefeller Foundation. There, he spent 13 years trying to
develop sustainable alternatives to the kind of farming
practiced by Nigerians and many other Africans.

Unlike the farmers in his Indian village who worked
the same fields for years, the Africans practiced what's
called "shifting" or "slash and burn" agriculture. They
cleared plots in the forests with a machete, then burned the
wood to provide nutrients for the soil. They raised corn,
yams, and casaba on their plots until the land simply gave
out. That tended to happen in just a few years, since the soil
in many parts of Africa is very old and very shallow, with
often just a foot of good soil above the rocky subsoil. In the
parlance of agriculturalists, the rooting depth is low. After

the plot was exhausted of nutrients, they'd leave it fallow for 15 or 20 years to recover. But as the population grew, they were running out of virgin forest to turn into farmland and were surrounded by old, unproductive plots. Lal's job was to help them restore fertility to the soil.

Lal began clearing areas for his test plots. His first effort was close to an utter failure. He plowed the land after completely clearing it—even pulled out tree stumps, which the local farmers didn't do—and built terraces to control water runoff. He planted maize, cowpeas, and rice, and everything seemed well. But the night before his board of directors was to view his work, 4 inches of rain fell in 2 hours. All his plots save one washed away, leaving behind gaping gullies in the soil. The one plot that survived had been heavily mulched with crop residue.

He asked his board of directors to postpone its visit, but he learned from the failure. He realized that disturbing the soil by plowing made it susceptible to erosion. Plowing also destroyed the structure of the soil—the internal architecture of sand and silt and clay created over the decades by earthworms and other organisms that allowed air, water, and nutrients to circulate. He decided to disturb the soil as little as possible when he prepared his next plots. He cleared some land by hand, some with a machine that cut the forest at ground level, and some that killed the vegetation with chemicals, leaving the roots intact. He sowed some plots with cover crops, which are typically grown not to eat or sell but to enrich and protect the soil during fallow periods. And he mulched some crops heavily, wanting to demonstrate how the chopped-up crop residue would protect the plots from being washed away by the rain and protect the precious carbon-rich topsoil.

His goal was to maintain the same carbon level in his plots as there was in the soil of the virgin forest, but he failed, over and over. There were complications he couldn't seem to overcome. In the heavily weathered old soils, the clay was as fine as powder and did not bind to the carbon as tightly as it did in other soils. Even though he didn't use plows, the machinery that drilled holes in the earth and dropped in the seeds compacted the earth and disturbed the soil structure. And though he mulched with crop residue, it never seemed to be enough to increase soil fertility—in the heat of central Africa, the residue decomposed so quickly that soil organisms hardly got a chance to eat it. And besides, the point of the whole exercise was to come up with methods of farming that the locals could easily replicate. But, as in his home village, they didn't want to mulch with crop residue when they could sell it in the marketplace or feed it to their animals.

"They didn't care about making their field richer next year," Lal says. "They couldn't afford to care about tomorrow."

Visitors passed through often to see Lal's efforts, and there was even a viewing platform so they could look over the entire 100 acres where his test plots were located. One day in 1982, a famous visiting scientist passed through and the institute sent him out to consult with Lal. They gazed over the fields together. The ground was as cracked as the varnish on an old tabletop, and it was just about as hard. And it was red, a sure sign that the carbon had leached out.

Then Lal led his visitor to the forest and turned up a spadeful of soil. There, it was dark and crumbly and crawling with earthworms. In the forest, the soil's carbon content was 2 to 3 percent. In Lal's plots, there was a scant ½ percent of carbon left.

"Where do you suppose the carbon has gone?" the visitor asked.

"I don't care where it's gone," Lal answered miserably. "I just want to put it back."

But the visitor was Roger Revelle, an oceanographer who was among the first to make the connection between the growing percentage of carbon dioxide and other gases in the atmosphere and changes in the earth's climate. With the chemist Hans Suess, he'd written an alarming paper for the journal *Tellus* in 1957 on this subject, in which he said human beings were conducting an unprecedented experiment with their environment by loosing so much carbon into the atmosphere. Lal hadn't read it. He was unaware of the connection between the loss of carbon he was witnessing in the soil and the steadily escalating measurements of carbon dioxide Revelle had been observing in the atmosphere.

"Roger was a great scientist—a giant," Lal says now.

"He explained that some of my lost carbon was going into the atmosphere and adding to the load of greenhouse gases. I had never understood the interaction between soil and climate, but at that point I stopped being so focused on soil alone."

Back at Ohio State in 1987, Lal found that, because of Revelle's work, both the USDA and the Environmental Protection Agency (EPA) had begun to pay attention to this interaction. With a few other scientists, he formed a working group on soil carbon and global warming, and they began trying to figure out two big issues. How much soil carbon had both the United States and the world already lost? And was it possible to put it back?

It's tempting to think that the loss of soil carbon is a relatively modern curse, the result of surging populations in poor countries and industrial farming in rich ones. But this is not the case. As soon as humans segued from a hunter–gatherer lifestyle to an agricultural one, they began to alter the natural balance of carbon dioxide in the soil and the atmosphere. Settled agriculture began in the world's great river valleys—those of the Tigris, Euphrates, Indus, and Yangtze rivers—some 10,000 to 13,000 years ago. By around 5000 BC, people began to develop simple tools to plant and harvest. The earliest of these were mere digging sticks, but by 2500 BC, people were using animals to pull plows in the Indus Valley.

Plowing seems so harmless and soothingly bucolic, especially when the plows are pulled by oxen or horses. But as Lal pointed out in a speech given in 2000, "nothing in nature repeatedly and regularly turns over the soil to the specified plow depth of 15 to 20 centimeters. Therefore, neither plants nor soil organisms have evolved or adapted to this drastic perturbation." Modern mechanized farming makes the problem even worse: The heavy machinery compacts the soil further, requiring deeper plowing to loosen the soil. As greater volumes of soil are churned up and exposed to the air, the soil carbon—which may have been lying in place under the soil line for hundreds or thousands of years—meets oxygen, combines with it to form CO_2, and departs for the upper atmosphere.

Animal husbandry also began upsetting the carbon balance. Before they were domesticated by humans, herds of ruminants roamed the great prairies, nibbling off the tops of the grasses and other plants and graciously dropping off loads of enriching manure in return. Fearful of predators, they clumped together tightly and never grazed in one place for too long. Humans affected a drastic change in the grazing patterns of these herds, though. Instead of continually drifting across the plains, the animals were either restricted to one area by fencing or they grazed freely under the protection of human herders and dogs. In fenced areas, they grazed right down to the bare ground. They often did that in spots under the watchful eye of herders, too—since they had no need to fear predators anymore, they'd loaf around in one place long enough to rip the plants' roots right out of the ground.

But allowing the animals to reduce grassy plains to bare ground halted the great biological process that had created vast underground stores of carbon in the first place: photosynthesis. Plants remove carbon dioxide from the air and, combined with sunlight, convert it to carbon sugars that the plant uses for energy. Not all the carbon is consumed by the plants. Some is stored in the soil as humus—Lal points out that "humus" and "human" share the same root word—a stable network of carbon molecules that can remain in the soil for centuries. There in the soil, the carbon confers many benefits. It makes the soil more fertile. It gives the soil a cakelike texture, structured with tiny air pockets. Soils rich in carbon buffer against both drought and flood: When there is rainfall, the soil absorbs and holds water instead of letting it puddle and run off. Healthy soil is also rich with tiny organisms—an amazing 6 billion in a tablespoon—that can disarm toxins and pollutants that soak into the soil through the rain. Lal believes farmers should be compensated not just for their crops; they should also be compensated for growing healthy soil because of its many environmental benefits.

No other natural process steadily removes such vast amounts of carbon dioxide from the atmosphere as photosynthesis, and no human scheme to remove it can do so on such a vast scale with any guarantee of safety or without great expense. Photosynthesis is the most essential natural process for life on our planet, as it regulates the steady cycling of life-giving carbon into our soil and creates that other gas on which so many of us depend: oxygen.

Lal and his colleagues developed a simple if crude method of estimating the amount of carbon lost from soils in the United States and the world. When I visited him at Plot 87, he gestured at a fringe of dark forest against one side of the test fields. "That forest is my baseline," he said. "When I calculate how much carbon has been lost from the soil in this plot and nearby areas, I compare it against the soil in the forest."

With funding from the EPA, the USDA, and the United States Department of Energy (DOE) and working with students and postdocs around the world, he compared the carbon in forested areas with that in cultivated areas. According to his calculations, Ohio has lost 50 percent of its soil carbon in the last 200 years. But in areas of the world where cultivation has been going on for millennia, soil carbon depletion is much higher—up to 80 percent or more. Altogether, the world's soils have lost up to 80 billion tons of carbon. Not all of it heads skyward—erosion has washed some of it into our waterways—but even now, land misuse accounts for 30 percent of the carbon emissions entering the atmosphere.

And the amount of carbon dioxide in the atmosphere has reached a truly staggering level. By 2013, scientists calculated that CO_2 had reached 400 parts per million (ppm) in the atmosphere—50 parts per million beyond the level that many experts think can reliably keep the climate stable for human life. Around the world, many clean-energy technologies are being devised and implemented to reduce the amount of CO_2 our modern lifestyles emit—from fossil fuels to wind, solar, biomass, and ocean-wave energy, and even, in one wild scheme, supplementing the power grid by salvaging the power of the body heat in crowds. And there are many strategies being used to decrease the amount of energy we consume, including bumping up the fuel efficiency of gas-powered vehicles and building homes and offices that generate more energy than they use.

However, none of these will actually reduce the legacy load of CO_2 already in the atmosphere. There are schemes afloat for doing that, but they're expensive—consider the EPA's plan to capture and inject atmospheric carbon into deep wells at a cost of $600 to $800 per ton. Not as sexy to policy makers, but free of cost, is Mother Nature's low-tech approach: photosynthesis and the buildup of carbon in the soil that naturally follows.

And therein lies our great green hope. To be sure, we must continue to cut back on fossil-fuel use and lead less energy-squandering lives. But we also have to extract excess carbon from the atmosphere by working with photosynthesis instead of against it. Farmers, ranchers, land managers, city planners, and even people with backyards have to make sure plants are growing vigorously, without large stretches of bare earth—photosynthesis can't happen on bare earth. We have to take care of the billions of microbes and fungi that interact with the plants' roots and turn carbon sugars into carbon-rich humus. And we have to protect that humus from erosion by wind, rain, unwise development, and other disturbances.

Lal says it can be done. The greatest opportunities are in the parts of the world where carbon has been most depleted by thousands of years of farming, in sub-Saharan Africa, south and central Asia, and Central America.

"The carbon in the soil is like a cup of water," Lal says. "We have drunk more than half of it, but we can put more water back in the cup. With good soil practices, we could reverse global warming."

When good land management practices create a ton of carbon in the soil, that represents slightly more than 3 tons of carbon dioxide removed from the atmosphere. Lal believes that 3 billion tons of carbon can be sequestered annually in the world's soils, reducing the concentration of carbon dioxide in the atmosphere by 3 ppm every year. But others with whom I spoke—especially as I got further and further from academia—are far more optimistic about the potential for change. This is still a new idea, they say, and science has barely nibbled at its edges.

By working with test plots around the world—in Ohio as well as Africa, India, Brazil, Costa Rica, Iceland, and Russia—Lal's center is looking for the perfect combination of land management practices in various climates and soil types that will remove carbon from the air and build it back up in the soil. He and his colleagues have figured out how to rebuild soil carbon in ecosystems across the globe, even in Nigeria, his early nemesis. They employ a variety of approaches, since the world comprises many microclimates and each has a different history of impact, human and otherwise. The one constant around the world is the importance of building political will. For various reasons, it's been hard for us to change.

Lal has written hundreds of papers and several books, including *The Potential of US Cropland to Sequester Carbon and Mitigate the Greenhouse Effect,* which made its way to president Bill Clinton and to America's delegation to the United Nations' Kyoto Protocol negotiations. Lal has spoken to Congress about the subject six times. In 2011 alone, seven international conferences addressed the connection between soil and climate. Still, Lal's ideas haven't sparked much follow-through among policy makers.

"Soil research is not attractive to politicians," Lal says. "I tell them about 25-year sustainability plans, but they only have a four-year span of attention."

But Lal's ideas and those of other land-use visionaries are sparking plenty of interest and action among those who take the long view. Today, we're experiencing an agrarian renaissance. An interest in wholesome, sustainably raised foods has caused an upsurge in demand, and the number of small farmers in the United States is growing for the first time since the Great Depression: Between 2002 and 2007, the number of small farms increased 4 percent. As these new, often college-educated farmers practice the kind of agriculture and animal husbandry approved of by their customers—reducing or eliminating the use of fertilizers, pesticides, herbicides, hormones, antibiotics, and other chemicals, as well as letting their animals graze on grass instead of stuffing them with food they didn't evolve to eat— many are surprised to find their soil changing. It's becoming blacker and richer with carbon. Some of these farmers don't care a whit about global warming; they're influenced by the industry-connected American Farm Bureau, which claims that 70 percent of farmers don't believe in human-induced climate change. But many other farmers are thrilled to find out that their humus is helping to keep excess carbon dioxide out of the atmosphere. They've become citizen scientists, testing new ways to "grow carbon," as well as entrepreneurs trying to figure out how they can get paid for this new crop.

The environmental community is also taking heed of the soil's potential to address climate change. Worldwatch Institute issued a 40-page report about the connection between soil and climate in 2010. The National Wildlife Federation has targeted global warming as the single great-est threat to wildlife and issued a report in 2011 on "future-friendly farming" that can mitigate climate change. The environmental community has been leery of embracing land-use management to combat global warming, worried that doing so might soften pressure on the energy and man-ufacturing sectors to reduce emissions. But the growing understanding of the link between global warming and soil carbon is revolutionizing the environmental movement.

It's revolutionizing *me* and the way I think about soil. I'm the granddaughter of farmers and the daughter of avid gardeners. I grew up against the aural backdrop of their discussions about their own and other people's gardens. There was never a car trip that didn't involve pulling to the curb to admire someone's bougainvillea or bottle brush. There was never a trip from one part of the state to another that didn't include several side trips to their favorite fruit stands (Patty's Perfect Peaches, are you still there?). No matter where they lived, my parents were never without well-tended flower beds, a large vegetable patch, and a compost pile. Even when she was in her early nineties, I saw my mother struggle up from her chair with dismay when a guest threw a tea bag in the trash. "We do it this way," she said, even though there wasn't really a "we" anymore, as my father had been dead for several years. She pulled away the tea bag string and pried out the staple, then put the tea bag in a white ceramic jug that she kept under the sink. She was still making compost for the 3-square-foot plot she kept in her senior apartment complex. When she was on her deathbed, silent for days, none of the family's attempts at engaging her in conversation worked, until my brother Dave exclaimed, "Mom, I just planted my tomatoes!" She raised up on her elbows and muttered, "Black cherry tomatoes?" It was a variety she'd become fascinated with after I'd brought a basket home from a farmers' market. That was the last thing she said. She died a few hours later.

So I was raised to appreciate soil and the people who

work with it. I first heard about Lal from a farmer named Abe Collins, who'd taken Lal's ideas and those of other scientists to transform his land and then became an evangelist for soil carbon. When I talk to Collins on the phone, he always sounds as if he can barely catch his breath. Part of this is purely physical; he's usually just run in from moving his cattle from one field to another or working on his fences. But part of it is excitement at the idea that he and others have stumbled upon something that really matters to the world, and that they'd better hurry up and get everyone to listen.

I would love to doubt global warming—or, more accurately, global climate change, because earth's atmospheric temperature has indeed risen 0.8°C since the industrial revolution, but that doesn't mean it's warmer everywhere. Instead, the weather is wackier everywhere, with a higher incidence of extreme weather events like downpours and droughts, floods and fires. I pine to be what's called a "climate denier," but the science won't let me. Scientists began tracking the buildup of carbon dioxide in the atmosphere decades ago. The measurements ticked steadily upward, but other, even more ominous data followed as the weather got warmer and weirder around the world. As environmentalist and author Bill McKibben wrote in a sobering 2012 article for *Rolling Stone* magazine, May of 2012 was "the warmest May on record for the Northern Hemisphere—the 327th consecutive month in which the temperature of the entire globe exceeded the 20th-century average, the odds of which occurring by simple chance were 3.7×10^{99} —a number considerably larger than the number of stars in the universe."

Still, it's a miserable conviction. I cringe when I read about polar bears drowning because their icy landscape has given way, or when I hear meteorologists predict seasons of increased hurricanes and pronounce each succeeding year the hottest since people began recording such things. I take no pleasure in an unseasonably warm winter day or an unseasonably cold spring, feeling as if the progression of the seasons has been shattered. I groan when global warming turns up as a story in my favorite magazines or a plot thread in a novel or movie. Because why think about it when huge policy changes are needed and policy makers seem incapable of making brave decisions? The actions of an ordinary person seem so paltry.

But for the first time since I read about global warming some 25 years ago, I feel hopeful. The soil will save us. I really believe that.

CHAPTER 2

THE MARRIAGE OF LIGHT AND DARK

Humans are not the first species to cause climate change. The original "polluters" were cyanobacteria—blue-green photosynthetic, aquatic bacteria that gave rise to plants—which began to change the balance of gases in our atmosphere some 2.9 billion years ago. Our planet was around 1.6 billion years old then. A mere pup.

Scientists are divided about the genesis of life on earth. Some believe extraterrestrial microbes caught a ride on a comet or asteroid that smashed into earth's surface, spilling the seeds of life onto a bleak surface comprising only rock and water. These went on to evolve into all the life forms we know today. Others, like Harvard's Martin Nowak, PhD, believe that life arose within the primordial soup, which contained minerals that reacted to one another and eventually formed chemical compounds. Some of these compounds changed in ways that made them stronger than others and, eventually, one or more produced an innovation that would change everything: Encoded in their compounds was a pattern that allowed them to self-replicate. At that moment, life—defined as something that changes and reproduces— began, and earth's early chemistry gave rise to biology.

No one is sure when those early life forms evolved or arrived. "All this happened so long ago that the molecular signatures are gone," Indiana biochemist Carl Bauer PhD, explained to me in one of several generous phone interviews. Scientists aren't even completely sure that the earliest life forms still exist, as one theory holds that early earth was smacked by a catastrophic event that wiped out the very first life forms. But most believe that the simple one-celled organisms called archaea (meaning "ancient") that live deep in the ocean near volcanic vents were the first, evolving in the intense heat of earth's past. The primordial soup was blisteringly hot.

Those were our first kin. Monkeys? We're sitting on each others' laps at one end of the spectrum of geologic time!

Could the archaea themselves have taken over the planet? The archaea feed on the swirl of chemicals around these deep-sea vents; however, that food source wasn't powerful enough to allow them to become large or complicated. But mutation was cooking up great variety within earth's first population, and these varied contenders for survival reproduced and spread. Bacteria evolved about 3.5 billion years ago and were also living in the ocean—a safe place, since there was not enough oxygen in the atmosphere to protect living things from the sun's killer ultraviolet light. At some point, a bacterium hit upon another innovation, one that would someday make earth unique in our chunk of the universe. Bobbing in the oceans only a few feet below the surface, competing with others of its kind for resources, it developed a mechanism that turned sunlight—a practically unlimited resource—into food. Thus began photosynthesis.

Back in 2001, Bauer's lab discovered that the first photosynthesizers were single-celled purple bacteria. They're still abundant around the earth—from the soil underfoot to ice-capped pools in the Antarctic to Soap Lake in Washington, where the water several feet down periodically turns the color of red wine. These pioneers developed a process for capturing light energy with pigment—in this case, a purple shade of chlorophyll, which is similar in structure to the heme pigment that gives our blood its characteristic red color and carries oxygen around our bodies—and spinning electrons from sulfur to power the conversion of solar energy into cellular energy. They and their descendants were called "photoautotrophs," meaning that they make their own food by absorbing energy from the sun.

For millions of years, this form of photosynthesis was the most advanced technology on the planet. The purple bacteria thrived, limited only by the amount of sunshine reaching them in the water. There were so many of them bobbing around that parts of our ancient seas would have looked purple. Then another organism called cyanobacteria tweaked the process again to momentous effect. This time, the cyanobacteria turned to water, the most plentiful molecule on earth, as a source for electrons, snatching away the hydrogen and liberating oxygen. "As soon as these photosynthesizers started using water as the source for their electrons, they were home free," says Washington University in St. Louis biologist Himadri Pakrasi. "They could grow anywhere on the planet because water is so plentiful. They took over."

Thus began the oxidation of earth, which would look as barren and inhospitable as Mars without the inadvertent ingenuity of these oxygen-producing photosynthesizers. Before they arrived on the scene, earth's atmosphere was a suffocating blend of ammonia, carbon dioxide, hydrogen sulfide, and methane, with a mere 2 percent pinch of oxygen. The cyanobacteria cranked it up to 21 percent over the millennia, the percentage where we comfortably reside today. Scientists can date the onset of this great oxidation because it raised red flags in the planet's geology: Since oxygen is so highly reactive, it began pulling electrons from the ferric iron (with only two electrons), leaving behind rust-colored stripes of soluble ferrous iron in the rocks (ferrous iron has three electrons). This is the same process that happens when metal rusts.

Still, even an oxidized planet would be inhospitable to life as we know it today. There would be nothing to eat on land, unless all life forms evolved to siphon up bacteria as food or to eat chemicals, as the archaea on the ocean floor did. It took the next great evolution in photosynthesis—pioneered by plants—to create an environment that provides beings like us, and every other animal, with both air and food.

Plants also had their origin in the seas, beginning some 1.6 billion years ago when an alga swallowed a cyanobacterium and harnessed its ability to create energy from sunlight. Like the cyanobacteria, these resulting photosynthetic algae later tweaked the operating system to their advantage. This time, they used as their predominant pigments green chlorophylls. Plants' leaves became earth's first solar panels, as the green chlorophylls absorbed light from the highly energetic blue and red wavelengths of sunlight. This ample source of solar energy made plants the first very successful multicellular organisms—beginning, most likely, with simple algae, which put together long chains of cells inside a protective filament—which were eventually able to accumulate the biomass to form tulips, potatoes, and giant redwoods.

In addition to water and sunlight, plants and cyanobacteria require carbon dioxide molecules, which plants absorb through holes called stomata on their leaves. After the plant captures sunlight, it rips apart the carbon dioxide molecules and gets rid of the oxygen. The prize is the carbon. Plants use the energy they've absorbed from sunlight to convert this carbon into high-energy sugars that feed the plant through its life cycle. Every cell in a plant contains chlorophyll and thus carries on photosynthesis. Even flowers—thought to be an adaptation of leaves—contain enough chlorophyll to carry on small amount of photosynthesis, although we can't see the green for all the gaudy pigments there to attract pollinating insects. Even the gnarly bark of trees that look about as green as tire tread conducts photosynthesis.

Although we humans fancy ourselves a clever species—and we are—we haven't, in all our cleverness, invented anything that compares with photosynthesis. The carbon sugars created though photosynthesis are the building blocks of life—they are the beginning of the food chain for just about everything on earth, as animals eat the plants and some animals eat other animals. Whether we humans dine on dill pickles or duck confit, we're eating these carbon sugars that the plants created from sunlight. Along with cyanobacteria, plants are also the beginning of the food chain in the seas. (In the seas, much of the burden of keeping everything else alive goes to the cyanobacteria, which ride the waves and convert sunshine into carbon sugars that are gobbled by zooplankton and other life forms.)

But here's the thing: *plants leak.* And in doing so, they support another world entirely: The world under our feet; the dark kingdom of which we're astoundingly unaware; the down under that may account for up to 95 percent of our planet's species diversity. This is the world of soil microorganisms. Dig up a teaspoonful of healthy soil from your garden or from a city park or from the weedy strip alongside a highway, and you're looking at something like 1 billion to 7 billion organisms, depending on the health of the soil. Scientists guess that as many as 75,000 species of bacteria could be in that teaspoon, along with 25,000 species of fungi, 1,000 species of protozoa, and 100 species of tiny worms called nematodes—and the count of microorganisms keeps rising, as scientists figure out better and better ways to look for them. When I was a kid, my favorite book was Dr. Seuss's

Horton Hears a Who!, in which Horton the Elephant is suddenly alerted by a tiny voice coming from a speck of dust that a microscopic village there called Whoville is in peril. I recall that the drawings in the book portrayed a modest town. The Whoville in that teaspoon of soil is more like Mexico City. Imagine how many microorganisms are in a cup of healthy soil. More than all the humans who have ever lived.

The plant world's relationship with this underground kingdom began as waves splashed them onto shore and they tried to make a living in the new terrain. By this time, the sun, wind, and rain had already prepared three of the basic building blocks for soil. From the smallest to the largest particles, they are clay, silt, and sand, all of them tiny bits broken from earth's stony surface. Bacteria and cyanobacteria had long since colonized land, along with another newly evolved organism: fungi, a life form that is neither plant nor animal but shares characteristics with both. These organisms had also been breaking down rock to get minerals, had already developed mutually beneficial relationships for cycling nutrients back and forth among them, and the fungi—at this point mostly the saprophytic kind, which feed on dead things—kept these new life-forms from smothering in their own organic waste. Their predators had also arrived, adding another layer of nutrient cycling to the mix. By the time rooted land plants evolved in freshwater ponds, this "soil food web," as microbiologist Elaine Ingham, PhD, calls it, was already functioning.

"This nutrient cycling had been going on for millions

of years before plants put down roots to hold themselves in place," says Ingham, who was the science director at the Rodale Institute in Pennsylvania from 2011 to 2013. "Plants didn't need to evolve a way to get nutrients from rocks, because a system was already in place. All they had to do was put out the right cakes and cookies—the right flavor of root exudates—to grow the organisms that make the enzymes that solubilize the nutrients they need from sand, silt, and clay."

So began one of life's great biological partnerships, a marvel of mutualism and coevolution. This partnership provided plants with enough nutrients to survive their harsh new environment without the protective lap of water, grow more complicated biomass, and spread their green across the land. And in exchange, the microorganisms received precious carbon sugars as well as protein and carbohydrates from the plant—truly, the elixir of life, the first superfood— and paved the way for the eventual evolution of animals.

Maybe the plants leaked carbon sugars from their earliest days and the bacteria got wind of this feast and made it their business to set up shop nearby. Or, maybe the plants began leaking carbon sugars in order to attract the microorganisms. Because this is never a one-way exchange in which the microorganisms leech off the hardworking plants and give nothing in return. Instead, plants and soil microorganisms have developed a sophisticated trading network over the millennia by which plants shunt up to 40 percent of their carbon sugars to their roots and microorganisms pay for these goodies by delivering a mix of minerals to the door like pizza deliverymen. Plants need these minerals to build mass and create enzymes for biological activity, and the creatures who eat them require the minerals to build healthy bodies. Australian ecologist Christine Jones calls this symbiosis "the very first carbon-trading scheme."

The partnership became even more complex and rewarding with the later evolution of mycorrhizae (from the Greek words *myco* for "fungi" and *rhizae* for "roots," and commonly called mycorrhizal fungi). By then, the most sophisticated plants were productive grasses—ones that produced grain heads—but they didn't have the energy resources to move much beyond that point. Fungi that possibly began as root parasites that pierced plant roots with tiny threadlike hyphae to suck out the cytoplasm "realized" (it's hard not to assign intent!) that siphoning out carbon sugars and depositing a payment in nutrients was better than killing or weakening the host plant. The plants not only survived the penetration, but also thrived since the hyphae of mycorrhizal fungi range far and wide and can connect whole communities of plants with nutrients. With this fungal root system in place, plants had more energy to put into reproduction and reaching toward the sun. Thus came shrubs and trees.

Fossils from the early Devonian period about 400 million years ago show preserved plant roots encasing ancient hyphae. And mycorrhizal fungi swap nutrients with plants in exactly the same fashion today. Scientists estimate that the roots of 80 percent of earth's plants are intertwined with these hidden helpmates.

But life underground is far more complicated than even that scenario, which is far more complex than anything I had ever imagined. I always thought plants absorbed nutrients through their roots without needing microscopic allies. If someone had told me that fungi were penetrating the roots of my beloved bee balms or daylilies, I'd have assumed that was a problem. How long have people known about these complex relationships and trading systems going on under our feet?

Really, not that long.

Anton van Leeuwenhoek was one of the first to see and describe bacteria, back in the 17th century. He was far from a trained scientist—he was a Dutch fabric merchant by trade, but had also worked as a surveyor, a wine assayer, and a city official in Delft. He kept interesting company, and was the trustee of the painter Johannes Vermeer. He had a lively mind and a talent for tinkering. Inspired by a popular book of the day showing images of bird wing feathers, bugs, and other natural objects under a compound microscope (one that uses more than one lens), he decided to try making his own. He wound up building more than 500 simple microscopes, some of which magnified over 200 times—impressive in comparison to the compound microscopes of his contemporaries, which only magnified 20 or 30 times. He was the first to see the invisible world of microorganisms. And he looked everywhere! In lakes, in drops of blood, in samples of his own feces, and in plaque taken from his teeth and from the teeth of two old men who never cleaned their teeth. From this "white matter, which is as thick as if 'twere batter," he made these observations:

"I then most always saw, with great wonder, that in the said matter there were many very little living animalcules, very prettily a-moving. The biggest sort . . . had a very strong and swift motion and shot through the water (or spittle) like a pike does through the water. . . . The second sort . . . oft-times spun round like a top . . . and these were far more in number." In the tooth scum from one of the dirty old men, he saw, "an unbelievably great company of living animalcules, a-swimming more nimbly than any I had ever seen up to this time. The biggest sort . . . bent their body into curves in going forwards. . . . Moreover, the other animalcules were in such enormous numbers, that all the water . . . seemed to be alive."

These sorts of observations didn't lead to an understanding of the role bacteria or other microorganisms play for a good many years. Most people assumed that the strange organisms that they found in the soil were harmful to plants. But in the late 1800s, scientists began looking a little closer. A group of German foresters eager to grow truffles persuaded German botanist Albert Bernhard Frank to study the woods and to help them figure out how to propagate the delicacy. As Frank dug up the forest soil, he discovered the silky cocoons of fungal hyphae near the roots of the trees. Some of these hyphae are only ⅟₆₀ the size of the plant roots; they lace together healthy soil—in a cubic foot, there can be 320 miles of hyphae—even if we don't notice them. Frank observed that the trees growing from the tangle of hyphae were healthy, and he began to suspect that the fungi were helping plants, not attacking them. He conducted an experiment in which he planted seeds in forest soil, some of which had been sterilized. The seeds in the sterile soil did poorly versus the natural forest soil that was rich in fungi and other organisms.

Although Frank's conclusions weren't universally embraced, other researchers went off to study soils throughout the world to see how widespread mycorrhizal fungi were. They found them everywhere, from the tropics to alpine regions, but not, significantly, in soils that humans had disrupted, like mines or other areas where the topsoil had been stripped away.

Scientists went on describing and defining soil microorganisms, but no one spent much time figuring out why they were there. It's possible they had a hard time finding funders for research into something you can't see and that most people fail to muster the proper enthusiasm for. It's also hard to conduct this research. Science typically pulls one piece out of a system for study, but the microorganisms in the soil are part of a complex system and can't really be understood in isolation. Some 99 percent of soil organisms can't be grown in a lab for study—perhaps because they require the in situ relationships to survive.

Understanding the microbial underworld had to wait until the 1980s, when biologist Dave Coleman urged his Natural Resource Ecology Lab (NREL) at Colorado State University to start trying to figure it out. He hired microbiologist Elaine Ingham as a postdoc, and she began studying the data. Weekly lab meetings went all day as she and the other members of the lab excitedly puzzled out the complex roles and relationships among the soil biology—not just bacteria and fungi, but a whole host of other organisms that are busily at work under our feet.

"No one had ever asked why these organisms are all there *together*," says Ingham, a woman with a friendly if determined face—she looks as I imagine those few people who stayed on and faced down the hardships of the American Dust Bowl's black blizzards in the 1930s might—and an almost inexhaustible enthusiasm for talking about soil organisms. She asks: "Why are protozoa in the soil? What would be the purpose? We know they eat bacteria, but what's so useful about eating bacteria? How does that help the plant in any way, shape, or form?"

The NREL ultimately figured out that it takes a village to nurture a plant. From agapanthus to azaleas, begonias to buddleias, cranesbills to coreopsis, throughout the entire botanical pantheon, when you look at a healthy plant you're seeing the productive output of a busy village down around the roots, making sure it gets everything it needs.

And what a fascinating world! As I learned more and more about our own dark side from Ingham and others, I felt dizzy standing on top of all that activity. When I was a kid, I used to love to lie on my back with my feet propped against the wall and imagine that I was in an upside-down world, where I'd walk on the ceilings and step over doorways to get into another room. Learning about life in the soil has been just as topsy-turvy. Most of us assume that all the action is on the ground, in the air, in the water, and that earth below the soil line is inert and lifeless—except, that is, for those plant roots.

But it's alive down there. Plant roots can plunge as far as 200 feet. Even some of the grasses that we grow in our parks and on our lawns, if they're healthy, have roots that can descend 15 feet—and every millimeter of those roots is thrumming with microorgasmic bustle. Microorganisms themselves have been found as far as 10 miles down in the soil. Oil companies have to be careful not to contaminate deep pockets of oil with these organisms, because they will happily feed on the oil.

Ingham's taxonomy of soil microorganisms includes five main categories: fungi, bacteria, mobile one-celled organisms called protozoa, tiny worms called nematodes, and microarthropods, which are related to crustaceans and insects. Along with the soil dwellers we can see with our eyes—earthworms, beetles, voles, and the like—they compose what she calls the soil food web. More complicated and less fragile than a food chain, these denizens of the underworld are connected to and reliant on one another in myriad ways—just as we denizens of the overworld are.

Gathered nearest the plant roots are the fungi and

bacteria, both of which line up like pigs at a trough to get their carbon sugars. They're so tightly clustered near the roots that they form an almost impenetrable boundary between the root and soil pathogens that lurk nearby, trying to attack it. The barricade is not only a passive one; fungi can actually throw their ropelike strands around an interloper—say a root-eating nematode—and strangle it. The plant is their sugar daddy, after all. It behooves the fungi and bacteria to protect it.

Likewise, it behooves the plant to keep the fungi and bacteria around, well fed, and increasing in numbers, as they bring the plant nutrients that it can't get any other way. Both fungi and bacteria secrete enzymes that liberate minerals from the clay, silt, and sand, as well as from stones and actual bedrock. Not just potassium, a mineral routinely added to crops in conventional farming, but also the wide array of other nutrients that the microbes as well as the plant and those that eat it need to flourish. Ingham says that when she was in grade school, the list of essential nutrients students studied numbered three. The number had grown to 12 by the time she was in high school, 18 by the time she was in graduate school, and it later reached 32. "The list is going to keep growing until we include everything on the periodic table," she says. "Everything's important. There is a reason for yttrium on this planet! We don't need much, but we probably need some."

Fungi deposit these minerals—maybe even yttrium— inside the wall of the plant root, but the minerals foraged and eaten by the bacteria require the participation of other members of the food web to reach the plant. The plant is finicky about how these minerals are presented. One could drop minuscule fragments of cobalt or sulfur—both on the long list of necessary nutrients—near the plant's roots, but to no avail; these nutrients must to be mediated by biology for the plant to be able to use them. Only when minerals are swallowed by bacteria, which are then swallowed by a protozoa or nematode or microarthropod out hunting near the plant's roots, then excreted by those larger organisms in the vicinity of the plant's roots—only *then* do the minerals assume a chemical form that the plant can use. At that point, the roots can absorb the nutrients through simple diffusion.

With two exceptions, plants get all the minerals they need from the soil, needing only microbial—not human—intervention. The air is the source for the two other hugely important nutrients. Plants snatch carbon from the air, all on their own. Nitrogen is another necessary airborne nutrient—our atmosphere is 78 percent nitrogen—but plants can't remove it from the air by themselves. Again, they need help from a microbial partner. Legume plants like alfalfa, clover, lupine, peas, beans, and locust trees attract a certain kind of bacterium that converts atmospheric nitrogen to a form that the plants can absorb. When the legume plant dies and decomposes, its nitrogen reserve disperses in the soil and becomes available to the entire local plant community.

So there are lots of pizza boys down there. Lots of

delivery people bringing Thai food or tamales (like the couple in the neighborhood where I now live in Portland, Oregon, who sell them from coolers on the backseat of their car). It reminds me a little of that scene in *Porgy and Bess* where vendors clog the streets selling strawberries, honey, and crab. Another analogy: The plant is kind of like a 1950s housewife, waiting for the Fuller Brush Man. But not just waiting passively; it summons the bacteria that tote the kind of mineral load the plant needs. Ingham's analogy is that the plant prepares a wide variety of cookies and cakes with its carbon sugars to lure the bacteria carrying just the right nutrients. The plant can both vary and increase its output of carbon sugars in order to beckon the necessary partner microorganisms.

These partner microorganisms are remarkably different from plant to plant (and even from location to location: in this case, think of the difference between the large felines of Africa and those in the United States). Mycorrhizal fungi work with many plants and, in fact, can shoot out their hyphae up to 250 yards and connect to different plants at once, sharing their mineral wares within a whole community of flora. But bacteria are more particular. Some bacteria, having evolved to eat a particular kind of carbon sugar, will only cluster around a few kinds of plants. They're not quite as particular as a creature like the sand verbena moth, which feeds exclusively on the sand verbena plant and is endangered wherever that plant is in short supply, but they do have a more specialized role in the underground ecosystem and are critical to the health of certain kinds of plants. This is the reason it's difficult to restore a degraded habitat: You might be able to replace the plants that were once there, or at least some of them; likely, the diversity of plants in a natural state is richer than we know. But will all of them grow in degraded soils where their partner microorganisms have been killed off? Unlikely.

The soil bacteria are also specialized for certain conditions, which are determined by factors such as temperature and the moisture level. Even though billions may be clustered around a plant's root, not all of them are working at the same time. When the temperature rises, or during droughts and floods, some of them slack off and others pick up the pace.

So, the soil microorganisms provide the plant with both food and protection from predators. Their third critical role is to control the underground flow of water and gases by building tiny structures in the soil called aggregates.

Bacteria form the tiniest aggregates by catching a piece of clay, silt, or sand and anchoring themselves there with carbon-rich glue, which they produce from the plant sugars. They do this initially to keep themselves from being carried away by water moving through the soil, just as the bacteria around our teeth create glue—scum and plaque—to cement themselves in place. As they glue more particles to themselves—another piece of silt or maybe a decayed bit of plant material—a tiny structure forms, which both protects them from other organisms that want to eat them and creates space for gases and water. Bacteria have many of the same needs that we do: They need to breathe oxygen and expire carbon dioxide. Without the open space in the soil created by these billions of awkwardly shaped aggregates, the bacteria would suffocate.

Then fungi come along and gather up some of the bacterial aggregates and make their own lopsided aggregates, hiding their reproductive parts inside to protect them from predatory microarthropods. In healthy soils, there are trillions of these aggregates piled on top of each other, each creating space for gases and water to pass through slowly. The aggregates form spaces that cup the water and make it available to all the life in the soil. Instead of thinking you're on hard, dead matter when you stand on soil, imagine instead that you're standing on something as porous and dynamic as a living coral reef.

When I used to dig in my backyard in Cleveland, my shovel often slammed into a layer of clay and remained stuck there in the soil, reverberating from the shock of my effort. I'd dig the clay up and it was pretty much like what you'd pound and knead in a pottery class. That's an example of soil with few microbial aggregates. Microscopically, clay particles have a rodlike shape. Without microorganisms to jumble up the rods and glue them into three-dimensional snowflakes, the clay particles press tightly against one another, forming a barrier for shovels as well as gases and water. That's why clay is a good water-resistant material for everything from pots to houses. Up to half of the world's people still live in houses that incorporate clay in the building materials.

Here's another analogy: Imagine the smaller mineral particles like clay and silt as sheets of paper. When they're flat, they form a dense, heavy pile. But if you ball up each sheet, the same amount of paper occupies a larger and airier space.

Soils with a lot of sand have a different problem. The large sand particles don't impede the flow of water, but there's so much space between them that they can't hang on to it, either. In this case, microbial and fungal aggregates become like tiny dams in the soil, each holding a precious reservoir of water. When there aren't puddles on the surface and obvious signs of moisture, we tend to think that the ground is dry. We call it dry land! But it's a water world down there beneath the surface. The microorganisms move around on the water films between and within the soil aggregates. Even though these organisms left the sea millions and millions of years ago, they still slip and slide their way through the soil on tiny waterways.

In fact, healthy soil that's rich in microorganisms and heavily studded with their aggregates holds water like a sponge, slowly releasing it to plants as well as to rivers and streams. Healthy soil is the best protection for crops during a drought, as well as the best protection against floods anywhere. The soil is the earth's first water purification system, too: The microorganisms will attack and purge water of its pollutants, eventually draining it into a stream or aquifer in a pure form.

People who extol the value of these healthy soils talk

about these benefits—the drought protection, the flood prevention, the purification of water—as ecosystem services. As if growing healthy food isn't important enough by itself! They've been adding another ecosystem service to the list in recent years, as fears about global warming mount: carbon sequestration.

We tend to think of the "greenhouse effect" as being a modern obsession, but scientists have been puzzling over the factors that control earth's temperature for centuries. French scientist Jean-Baptiste Joseph Fourier noted in the early 1800s that the temperature of earth "should be a little below what would be observed in the polar regions," since it seems that the heat we receive from the sun should escape into space. Based on his studies, he suggested that our atmosphere provides an insulating blanket which keeps us warm. Later that century, John Tyndall conducted experiments at the Royal Institution of Great Britain showing that various atmospheric gases—including water vapor and carbon dioxide, now known to be the two most significant greenhouse gases—could absorb and emit enough radiant heat to control the planet's temperature. In 1896, Swedish scientist Svante Arrhenius published a paper showing that changing atmospheric concentrations of carbon dioxide might explain earth's cycle of ice ages and temperate periods, and noted that humans might have an influence on this concentration through the burning of coal. British engineer Guy Stewart Callendar built upon this idea in 1938, pointing out the links among fossil fuel combustion, rising CO_2 levels, and rises in world temperature.

Even if the burning of fossil fuels wafted more carbon dioxide into the atmosphere, most scientists assumed the seas would absorb this excess. But in 1959, Swedish scientists Bert Bolin and Erik Eriksson published a paper showing that while CO_2 was absorbed in the upper layer of the ocean, most of it drifted back into the atmosphere before sinking safely into deep water. In 1958, Harry Wexler, chief of scientific services at the U.S. Weather Bureau secured funding for a permanent station at the Mauna Loa Observatory in Hawaii to monitor the gas and tapped a young research chemist named Charles Keeling for the job. Keeling had discovered an accurate way to measure atmospheric carbon dioxide and pegged it at 310 parts per million. By the time he died in 2005, the reading had risen to 380 parts per million. By 2013, the count was 400 ppm.

Even if we miraculously stopped using fossil fuels right now—and we have so far lacked the courage and vision to make meaningful progress—we would still have this shroud of carbon dioxide hanging over our heads. This legacy load, as it's called, would eventually dissipate, but over thousands of years—not soon enough to save the planet from rapid warming. A handful of scientists and entrepreneurs have geeked-out plans to remove this legacy load from the skies, but—at least so far—their ideas either have frightening caveats attached or are so costly that they'd never get funding.

But we are already living within a massive biological machine that can tackle the legacy load. Nothing that we think of as our world would exist without this machine. We've been unwittingly hampering its work for millennia— more on that in the next chapter—but beyond the reach of our eyes and ears, it is always at work, removing carbon dioxide from the air and converting it to a precious resource. And this machine does its work for *free*.

When the fungi and bacteria gobble up the carbon sugars inside or near the plant root, the carbon doesn't just disappear. It becomes part of their bodies. Fungal hyphae snake that carbon throughout the soil as if they were railroad tracks; when they die, that far-reaching network of carbon stays in the soil to be nibbled at by other creatures. When other microorganisms eat the fungi and bacteria, they incorporate the carbon into their own bodies. Even digestion leaves deposits of carbon studded throughout the soil when the fungi and bacteria excrete carbon sugars in their waste, and still smaller creatures snatch it up and eat it. The carbon keeps cycling through the soil food web, and each time it's eaten and excreted it emerges in a more concentrated form. The process of decomposition is one in which the soil organisms keep creating longer and more complicated carbon chains. Thus, the carbon sugars in the simple syrup that the plant created from sunlight ultimately is bound into a chain with maybe 10,000 other carbon atoms, which are themselves linked to hydrogen, oxygen, and other nutrients. As the carbon chains grow bigger, all that carbon keeps making the soil darker.

The word for these carbon chains? "Organic." It's a

word that's become very fuzzy as marketers have applied it to everything from peaches to frozen pizza to cosmetics. For hundreds of years, soil chemists used the word "organic" to refer to compounds with carbon chains, all of which contain the energy that plants create from sunlight. The word only became a synonym for healthful and natural food when Jerome Irving Rodale—the founder of the Rodale Institute—applied it in the 1940s to the kinds of nutritious food he sought for better health. He was convinced that only the agriculture that worked with nature to create these carbon-rich soils could produce good food.

As one soil organism after another eats the carbon sugars and poops them out, the carbon chains become more recalcitrant—meaning, they're more and more difficult to break down further. At the end of the process emerges what's called humin, which is just carbon, hydrogen, and a little oxygen. There's almost nothing left to eat there, and that carbon can be locked into the soil for centuries. But when soil scientists talk about soil organic matter or humus, they're not just talking about humin. And they're not referring to the bags of stuff sold at the garden center. "Soil organic matter is not just one thing," says Kristine Nichols, a soil microbiologist with the USDA's Northern Great Plains Research Laboratory in North Dakota. "It's thousands to millions of things. It's the simple sugars, it's the bacterial cells, it's the waste produced by the bacteria and fungi and other organisms. What we call 'humus' is a broad spectrum of molecules."

Soil microorganisms breathe, just as we do, and they exhale carbon dioxide. Of all the carbon that goes into the soil—from the carbon sugars and the plant debris and from all the microorganisms that decompose both—only a tiny percentage is locked semi-permanently into the soil as humin. Nichols suggests that it could be as low as 1 to 10 percent. Still, it will remain in the soil for decades or centuries—even millennia—instead of cycling back into the atmosphere in months.

So back to Rattan Lal's idea that we can reverse global warming by locking carbon into the soil this way. Wouldn't we have to put an immense amount of carbon into the soil for that 1 to 10 percent to make much of a dent on the legacy load of greenhouse gases in the atmosphere?

How is that possible, when everything humans have done to the land in the last 10,000 years has stripped carbon from the soil and slowed the process of putting it there?

CHAPTER 3

SEND IN THE COWS

A cow with horns like a wide unstrung bow waved a sharp tip at a black-faced goat that had darted in front of her to grab a mouthful of grass. As if in a huff, she then swung her big gray head to the east and began to strut away from the herd, drawing a line of cows behind her.

Quickly, Soka stopped our conversation and jogged ahead of the breakaway bovines. He didn't wave a stick or throw stones or even shout, as herders often do. This wedge of herbivores—500 cows plus 700 sheep and goats— is managed differently from any in this part of Zimbabwe, differently even from how Soka's great-great-great-great- great-grandfather would have managed a herd. Creature- to-creature courtesy prevails. Waving sticks and shouting would stress the animals and make them less likely to flourish. So Soka just stood in front of the wayward cows, arms akimbo, his heavy jumpsuit a lean column of green in the golden winter landscape of southern Africa. The cows stopped in front of him and masticated thoughtfully, their dark eyes fixed upon him, then ambled back to the herd.

Soka knelt again and bent back the grass, some of it so dry that it snapped noisily. "We move and move and move the herd," he said with his commendable accumulation of English words. "They eat and eat, but not too much."

I nodded.

He spun around and pointed to a naked patch of ground behind him, so hard and dry that its surface was as smooth as a dusty clay pot. The cows had missed this piece. "When water falls here, it goes *whoosh*!" He flung his arms out at the surrounding savanna, his face a picture of feigned distress at the thought of that water running off the surface. Then he pointed to an area that the herd had passed over, nudged along with hisses by him and the other herders. Some of the grass had been eaten and the rest was trampled to the ground. Piles of manure steamed between the stalks. In the bits of bare ground between the tufts of grass, the animals' hooves had gouged little half-moon dents in the soil. Soka smiled. "Here, water goes down and down. Land heals."

I already knew most of this in theory, but enjoyed the tutorial from someone who every day put to the test Allan Savory's approach to healing the land. Later that evening, when I was back at the Africa Centre for Holistic Management near Victoria Falls, sitting at a table with people who were there either to learn or to teach the Savory method, I turned to Savory himself and told him about Soka's attempt to enlighten me.

Savory laughed and fingered his short white beard. "I

brought Zimbabwe's minister of water out here and set him up under a tree near one of the pools of water in the savanna to talk to one of our senior herders," he said. "It was interesting to see this educated minister being schooled by a totally illiterate herder, who was telling him in his own language that it was the hooves of the cattle that had produced this pool of water. He finally convinced him. We'll start healing the rivers of this country if we ever have normal political times here."

In all the discussions of agriculture and land use, it may be hard to find anyone as controversial and iconoclastic as Savory. So much so that I was actually taken aback by his physical appearance when I first met him. He's a slight man, and I was expecting a giant. I had arrived at the center 2 days before I went walking with the herders, after I'd spent a frustrating 3 hours at the Victoria Falls airport. Another foreigner named Kris was arriving later that afternoon, and this had so confused things at the center that no one was waiting for me. A savvy traveler friend back in the United States had warned me not to tell the Zimbabwean airport officials that I was a journalist—"They might make trouble for you"—but as planeload after planeload of tourists arrived and paraded away with tour operators, I had started to wonder if dictator Robert Mugabe's men had guessed my secret.

Someone from the center finally arrived and installed me in a small dormitory built for visiting students, where it was so cozy and dark that I was tempted to stay inside and catch up on some sleep. But I was in Africa for the first time in my life—Africa!—and had to explore, even though my driver told me that Savory couldn't meet me for a few hours. I put on my straw hat and set off down one of the dirt paths leading away from the center, stopping when the driver shouted after me. "Don't go too far!" he said. "We are in the jungle, and there are animals all around."

"Which ones?"

He held up four fingers. "All of the Big Five except rhinos. Elephants, lions, leopards, and buffalo."

So I wandered close to the center for a while, admiring the elegant local architecture. There were two large buildings called rondavels at the center of the campus, one a dining room and one a classroom. The walls were built of local stone set into a daub made from—Savory later told me—soil that termites mix with their own saliva, taken from the huge coned nests out in the savanna. Each building shouldered a high, gracefully thatched roof made from grasses harvested nearby that had weathered to silver and were cut in scallops along the bottom, with another decorative cap of grasses at the very top. Then I wandered slightly farther, past another rondavel under construction, its thatched roof still golden. I learned later that lions sometimes lurk near the under-construction rondavel, that I might have wandered a bit too far.

I finally settled at a thatched patio with a few of the center's other guests. The sunlight had started its afternoon slant and we were musing about wine, which the center sold on an honor system from the refrigerator in the dining room. Then a couple approached the table. I recognized Savory's beautiful wife, Jody Butterfield, from her photo on the jacket of their book, *Holistic Management: A New Framework for Decision Making*, then realized that the man with her had to be Savory himself. He carried a long walking stick, and my quick impression was of a desert patriarch in khakis. While everyone else had their legs covered, he wore shorts, which exposed thin brown legs and shoeless feet. I had felt the sharp prickling of the dried grasses around the center even through my sneakers, and I asked about his bare feet.

"I read the land with my feet," he said with an amused smile. "When I'm in New Mexico (he and Butterfield live half of each year in the States), I run barefoot on gravel to keep my feet from getting soft." He told me that his wife sometimes hears weird scratching sounds when he gets into bed at night, which usually turn out to be caused by thorns stuck in his feet snagging the sheets. He doesn't feel them.

Savory was born in 1936 into an Africa that he still pines for—an Africa where he could hear drumming at night and large animals crashing through the bush. Zimbabwe was still Southern Rhodesia then, a British colony dominated by whites and named after the South African politician and businessman Cecil Rhodes. But during his youth, Savory was less interested in the country's politics than in its wild places. His father was a civil engineer who often had to career into the wilderness to inspect dams or other projects. His uncle had a ranch where the young Savory spent as much time as possible, riding horses and hunting. Savory kept up with these passions even when he was sent to a boarding school called Plumtree, which was a feeder school for Britain's Royal Military Academy. Savory was fascinated by war—World War II raged during much of his youth—but still, he was a tough fit for the school. In defiance of rules, he brought his guns and sneaked out to hunt as often as he could, taking his catches to a nearby village to be cooked and shared. "I was very irreverent," he told me. "I kept my ammo in a hollowed out Bible on my bookcase. I knew no one would look for it there."

Savory was no less scrappy in college, but the difference between him and the academic environment was now more philosophical. He majored in botany and zoology, and his teachers constantly chided him because he'd raise questions about animals in his plant classes and plants in his animal classes. For his part, he couldn't understand how one could have a meaningful discussion of plants while excluding the impact that animal life had on them. Likewise, how could one understand animals without knowledge of the flora that made up the floor, walls, and often the ceilings of their world? "I couldn't get a discussion going about both at the same time," he said. "I was very discouraged with the compartmentalization of academic life."

He was also dismayed by what he saw as ignorance among his lecturers and had no problem challenging them. When one visiting lecturer told a class that crocodiles have flaps behind their ears and the musculature to move them but never did, Savory pointed out that his pet crocs—he not only had them at the university but even later, when he was in the Rhodesian army—agitated their flaps when he angered them. And Savory was irritated when scientists spoke of experiments done in laboratories with plants dug up from the wild. "The moment you dig it up and remove it from its environment, it's a different plant," he told me.

As soon as Savory finished college, he skipped all the graduation festivities and headed back into his beloved bush. At the age of 20, he became a research biologist and game warden for the Northern Rhodesian Game and Tsetse Control Department, in the region that is now the nation of Zambia.

Quickly, he said, "I began to realize that all I loved was doomed." Rapid desertification encroached upon the vast grassland savannas and threatened the habitat of the animals he was hired to protect; their numbers were thinning. Like everyone else then *and* now, Savory assumed that the land was degrading because too many cattle were also grazing the savannas. "I'm told you can go back to ancient Hebrew texts and see them blaming the nomads for causing the desert with their animals," he told me.

And it is certainly true that long-ago humans were hard on their environments. Many of us have dewy-eyed notions that people living before the 200 years of the modern industrial era lived in harmony with the land and left a lighter footprint than we do. However, research shows over and over that this was not the case. Their cumulative effect was lighter than ours because there were relatively few of them, but "their personal footprint per person as measured by the land use metric was actually much higher" says scientist William Ruddiman, PhD, author of *Plows, Plagues and Petroleum: How Humans Took Control of Climate.*

True, premodern people weren't driving cars and tearing the tops off of mountains to mine coal for electric power, but they burned forests to create pastureland and cropland, and they ripped open the soil with increasingly damaging plows to plant their crops. They didn't pave paradise, but there was so much of it that they could afford to ruin one plot of ground and just move on to the next green space. Of the total global deforestation that's taken place over the millennia, 75 percent occurred before 1850.

Humans began to affect climate with their carbon-releasing activities long before the modern era. Normally, greenhouse gases in the atmosphere fluctuate. Ice cores show that interglaciations, or ice ages, driven by a natural drop in greenhouse gases, occur roughly every 10,000 years. But the onset of human agriculture 10,000 years ago began to add new amounts of carbon dioxide to the atmosphere. Archaeological data shows that populations exploded and deforestation became rampant in Europe and China about 8,000 years ago, with an attendant belching of greenhouse gases—not just carbon wafting up from burning forests and degraded soils, but also methane from wetlands, irrigated rice fields, and livestock. Ruddiman theorizes that early humans increased greenhouse gases so much that we avoided an ice age that should have arrived perhaps 2,000 years ago. Without human activity over the ages, our atmosphere would contain about 245 parts per million, instead of the 400 ppm reached in 2013.

Early humans were so hard on the land that they created huge deserts in areas where humans have lived the longest, Savory says. These include the Sahara, a China-size desert in northern Africa, and the Tihama in Saudi Arabia and Yemen. Savory likes to point out that in the 5th century BC, Herodotus described Libya as having rich soils and an abundance of springs to feed its large population. Now, Libya is largely desert.

Much of the blame for desertification has been placed on pastoralists and their animals. Savory had no fondness for cattle or ranchers early in his career, and famously said in an oft-quoted remark, "Let's shoot every damn cow and bloody rancher that stands in the way" of the effort to reverse desertification.

Still, he spent a lot of time in the bush and what he was seeing there didn't always jibe with the cow-hating dogma. Ranchers stopped grazing their cattle in areas where the tsetse fly was rampant, but even in these wild swatches—where the elephants and zebras and big cats roamed without competition from cattle—the land continued to degrade. Game wardens typically set fire to the ranges to get rid of the old dried grasses—and indeed, there would soon be a flush of green among the blackened stubble—but even though the other scientists believed that fire would bring fresh grasses and the game back, Savory noticed that it increased the patches of bare ground between the plants. "It was true that herders had been degrading the land thousands of years," he said. "But a century of modern land management was just intensifying the degradation."

Range management thinking, then and now, held that degraded land would heal itself if the animals were removed and the land were allowed to rest. But over and over, Savory saw that rest did not heal his savannas. For instance, Zimbabwe's method of battling tsetse fly outbreaks was to kill all the game animals over large areas with the intention of removing the flies' blood meals and starving them. The practice reduced the tsetse fly outbreaks, but Savory noted that the land did not heal during these periods of time when all the animals were gone. Instead, it grew ever more degraded.

In yet another area—a game reserve called the Tuli Circle, near the Botswana border—Savory and other wildlife biologists saw the number of animals wax and then wane dramatically as overpopulation triggered mass starvation. They expected to see the land recover, but again it continued to deteriorate, even with fewer animals. Many of the other scientists blamed drought, but Savory noted in a research paper that they'd actually had a high-volume rainy season that year. He concluded that once land passes a certain point of degradation, it cannot heal—a conclusion he now believes is utterly wrong.

He continued to believe that drought and too many cattle caused desertification until he visited northern Europe for the first time. He observed Scottish lands that had been harshly overgrazed but then healed when rested. Even areas with low annual precipitation were able to heal. In fact, this is what I found in my backyard in Cleveland every time someone dug it up trying to fix the drainage problem that kept flooding my garage: If I didn't plant something, nature would quickly raise a bountiful and vibrant crop of weeds on the exposed soil.

How could prolonged rest heal lands in northern Europe and in the eastern United States but make the Rhodesian savannas more degraded? Savory finally realized that these environments were vastly different on what he came to call the brittleness scale, meaning whether the vegetation routinely dried out so much that it would snap in your hand. My backyard in Cleveland was right up against the far end of the non-brittle scale, because we had steady precipitation—either rain or snow—throughout the year. When I first moved to the eastern United States from my arid stretch of the Sacramento Valley, the moisture and especially the humidity were a shock. At first, I felt almost claustrophobic, as if I were imprisoned in a steam bath.

Unlike Cleveland and northern Europe, much of southern Africa has rainy seasons followed by long dry seasons. Savory surmised that the annual *amount* of precipitation wasn't the key factor in whether the land could heal all on its own, with rest; instead, the *distribution* of moisture throughout the year was critical. Bare land just can't heal when it's all dried out. It even develops a hard surface—like the bare clay-pot patch that Soka pointed out to me—that actually repels water. The vegetation in these brittle areas dies but does not decompose, which is a biological process carried out by microorganisms. These microorganisms die or go dormant when the rains stop. During these long dry periods, the vegetation instead undergoes a chemical process in which it oxidizes or just weathers and takes a long time to break down. The evidence for that had been in front of me on my first day at the center: the thatched roofs on the rondavels that had turned gray and could remain rigid and unyielding to the elements for decades. When left on the savannas, these dry grasses become a tough, durable screen that keeps the sunlight from reaching the soil and stimulating the growth of new plants when the rain returns.

But Africa's savannas had been rich environments for wildlife in the distant past. If they hadn't been able to heal themselves from fire, occasional overpopulation, and other disasters, what had healed them? Savory's earliest conceptualization of nature—that there was a complex interplay between plants and animals and soil that science erroneously picked apart—carried him toward new thinking. He guessed that pastoralists over the millennia were unwittingly subverting some natural process that had healed the grasslands, and so they were causing it to degrade.

Savory's musings about the savannas were interrupted

when Rhodesia erupted in civil war in the 1960s, as various black guerilla groups fought against the racist Ian Smith government. Savory was called into active duty in the Rhodesian army, and, given his experience in the bush, he was put in charge of a tracker combat unit that followed the movements of the guerillas. Where he had once observed the land closely to track problem animals or poachers, his observation skills now sharpened considerably, as his own and his comrades' lives depended on his ability to read the land. "I was observing the ground at the level that scientists usually don't," he told me. "You're going to get shot if you miss a bent blade of grass. At night, you're lying in the bush and you can't light a fire, because you're fighting a war. You've got all night to think, because you don't sleep very well."

Savory adopted the practice of walking barefoot through the bush then, not only because it made him more aware of the ground beneath his feet, but also because he didn't want to give his unit's movement away to barefoot guerillas who could easily pick up the imprint of a boot. All the men in his unit went barefoot too.

In the course of tracking the guerillas, Savory and his unit passed over wild grasslands where animals behaved as they had for millions of years, as well as through game reserves, farms, and ranches. He quickly found that the hardest places to pick up the track were areas where there were lots of herding animals behaving as they had before humans interfered with nature; there, the grass was lush and growing. In areas where people were managing the animals, whether cattle on ranches or elephants in game reserves where they were protected from predators, the land was poor and the grass sparse. Savory realized that animals in their natural state had a positive effect on brittle landscapes. In fact, it seemed that either removing them from the land or changing their ancient behaviors *caused* the degradation.

In the past, the savannas had been home to vast herds of grazing animals such as antelopes, buffalo, elephants, zebras, and many more. What was so very different about the behaviors of these herds? Savory recalled his observations of the many herds of wild elephants he'd tracked over the years. When they slowed down to eat in the grasslands, they spread out slightly, but not very far, because they were afraid of the lions, hyenas, and other predators that hunted them in packs. As they fed, they dropped dung and nitrogen-rich urine on the ground, feeding both the plants and the microorganisms in the soil. When they moved on as a herd, they stayed close together—again, to avoid getting picked off by packs of predators that didn't dare approach a herd—and trampled all the vegetation in their path. Savory realized that this trampling actually benefitted the soil. Grasses don't shed their leaves as many other plants do, and the trampling forced both living plant material and dead grass onto the soil surface—the dead grasses didn't get a chance to oxidize, weather, and block sunlight from seedlings. This plant "litter" protected the soil from erosion, shielded soil moisture from evaporation, and fed the soil organisms. The animals' hooves also beat up the surface of the bare soil, allowing it to take in seeds and moisture, in a manner very similar to the way gardeners break up the soil surface to prepare a flower bed. Humans unintentionally changed the way the herds impacted the grasslands when they domesticated them. The herds no longer needed to clump tightly together and keep moving to protect themselves from predators; there were now fences and vigilant herders to protect them. The animals became sedentary and spread out over the landscape, becoming the picture-postcard bucolic scene that we so love today. Savory realized that the land degraded because most people didn't understand the vital connection between plants, animals, and soil.

A trip to North America further convinced Savory that animal impact could heal the land. He visited arid national parks where cattle had been kept away for decades and where there were so few wild animals that there was literally no animal impact. Those lands continued to degrade, so much so that some people concluded that they were meant to be that way.

Savory doesn't claim that he was the first to notice the connection between large numbers of herding animals and soil health—folk wisdom from various countries held that animals could improve the land with their hooves. He recalled old ranchers telling him that one had to "hammer veldt to sweeten it"—meaning, to pound the land with hooves. So he was sure animal impact was needed to heal the land, but what kind of impact?

When white settlers arrived, the savannas had been bountiful with vast herds of buffalo, antelope, zebras, and the like that far outnumbered the domestic herds that arrived later. Savory began to wonder if the problem was not the number of animals on the land, but rather the amount of time they spent there. Again, conventional thinking held that animal impact was bad, so no one had thought of conducting such heretical research. He tried doing his own research by lobbing paint bombs at wild elephants to distinguish one herd from another so that he could measure exactly how long they stayed in an area before lumbering off together. This proved an impossible task, and he couldn't round up anyone else to help him.

Then Savory heard about a savvy botanist and farmer in South Africa named John Acocks who claimed to be grazing cattle in such a way that they made the land healthy instead of ruining it, so he flew down to meet him. Acocks explained that if cattle were turned loose in a large field to graze wherever they wanted, they would pick their favorite grass and graze it down to the ground, then move on to their next favorite and graze it down to the ground, and so on. The prevailing wisdom at the time held that this was an essentially good practice, as it allowed the cattle to select the nutrients they needed. But Acocks knew that this was a surefire way to ruin the land, because the cattle would over-graze the desirable grass types until their roots died and the land became bare. Instead, he confined his cattle to small areas so that they would graze the plants evenly and not too much, then moved them to another small field to begin again. Savory examined Acocks's field with great interest, noting how the cattle trampled plant matter into the soil, how their hooves had pocked the surface so that water seeped in instead of running off, how new plants were growing around the roots of the old ones. He saw that it might be possible to manage domestic cattle so that they replicated the salutary effect of wild herding animals. The amount of time they grazed the land seemed to be a critical factor.

Back in Rhodesia again, Savory consulted a book that had been sitting on a shelf, ignored, for years. French scientist André Voisin's findings were widely published 60 years ago but largely ignored in Africa, Australia, and the United States, which hold great sway in international range management science. Savory himself had never been terribly interested in the man's work, as he assumed that insights from fecund France wouldn't have much bearing on the dry African savannas.

But in *Grass Productivity,* Voisin explained that overgrazing was determined not by the number of animals on a plot of land, but on how long they stayed there. Acocks hadn't let his cattle stay in a pasture long enough to eat the grass down to the soil line, and he didn't return them to the pasture until the grass had recovered. This not only kept the land from turning into desert but actually improved the grasses and soil. When cattle or other grazers bite and pull at the grass, the plants struggle to regrow their leaves. They suck carbon sugars back up from their roots and root hairs, causing some of them to die and leave behind a web of carbon to decompose in the soil. As the grasses rebuild their stalks and leaves, they ramp up their production of carbon sugars again, shooting some back down to their root system for new growth. "When you graze and then let the plants recover, they pulse carbon and moisture into the soil," Savory explained. "Trees don't do the same thing. That's why grasslands are so important to carbon cycling."

Acocks had come close to having his cattle replicate the grazing patterns of ancient herds. Because these herds bunched tightly together to protect themselves from predators, they quickly covered a small area—similar to the size of Acocks's pastures—with urine and dung. No animal wants to feed on its own waste, so the herd kept lumbering forward instead of overgrazing that same piece of ground. When they reached a new stretch of land, they'd graze for a nutritionally balanced diet, eating some plants and leaving others, but generally leaving the land in good condition with their hoofprints, dung, and nitrogen-rich urine. By the time their dung had decomposed and they were ready to return to the land, it would have regenerated. Years and years of this behavior created carbon-rich soil filled with aggregates that allowed rainfall to quickly penetrate the soil and remain there, even through the dry periods. Savory calls this "effective rainfall," meaning that it percolates deeply into the soil and nourishes a wide range of living things over a long period of time. The rain doesn't run off and erode the soil, as it does in degraded landscapes, where there can be floods right after a heavy rain because the land can't absorb the water and then, the following week, parched soil and drought. If there isn't effective rainfall, Savory argues, then total rainfall doesn't matter much.

By the mid-1960s, Savory flipped his political alliances; instead of tracking black guerillas, he was now the leading voice of white opposition in parliament to the racist Ian Smith government. His interest in politics had stemmed initially from his conviction that environmentalists needed to get into politics. But just as he didn't believe in separating the science of plants from the science of animals, he couldn't separate what was happening on the land from the fate of people who lived on the land. "I wore many hats at that time," Savory told me. "But whatever hat I was wearing, I was always interested in poor land leading to the loss of wildlife, to poverty and violence, to the abuse of women and children, to political upheaval. In my mind, it was always the same thing."

He had also come to the conclusion that professional range managers were doing more harm than good to the land. He had battled the practice of controlled burns from within the Northern Rhodesian game department, arguing against the common belief that wildfires had created the wide, grassy savannas and pointing instead to the innumerable migrating animals that coevolved with the grasses. He argued that even if the burns created new growth in adult plants, they suppressed the number of new plants and created wider and wider spaces between the vegetation. Such loss of biomass, whether caused by fire or overgrazing, was difficult to reverse and he declared it the beginning of the process leading to desertification. He tried to publish scientific papers on the subject, but couldn't get them through the peer-review process. He proposed putting together a committee of young, bright scientists who would travel the world for a few years to see how people in other countries were managing their grasslands. Then he quit his job at the game department, convinced that it was just too hard to get people to listen to an argument that ran counter to everything they had ever learned. He no longer wanted to be a scientist employed by others, as he felt that bureaucracies and their dogmas hampered creative thinking. He became an independent researcher and consultant with a raft of clients, as well as a rancher himself.

Despite Savory's iconoclasm and occasional pariah status—or maybe because of it—farmers and ranchers began to ask for his help. One day an old ranching couple turned up at his front door. They had been following the advice of all the other experts, and they could see that their land was still deteriorating. Even Savory's infamous statement about wanting to shoot the bloody cows and ranchers that he once thought were causing desertification hadn't deterred them. "I told them that I didn't know they loved their land as much as I did," Savory told me. "I said, 'I'll help you, provided you understand that I have no answers for you. We'll have to work together and see what we can find out.'"

Even highly successful ranchers sought his perspective.

Rhodesia's conservation service gave an award every year for the best-managed veldt, and one of the winners contacted Savory to ask, "Is my land as good as the authorities say it is?" Savory asked the man's land manager to take him to the best area on the property and leave him for a few hours. "It was a sea of grass waving in the wind," Savory recalled. "It looked beautiful. It had rightly won the trophy."

But Savory dropped to the ground to do his observations, just as he had when he was tracking guerillas during the war. This time, he was measuring the bare space between individual grass plants. The other experts at the time measured desertification by the land's overall loss of productivity, but Savory's definition was becoming more nuanced. His included lack of biomass, which he found in the bare ground between plants. He also found that the roots of the mature plants were sticking a half inch into the air. "Obviously, they don't grow into the air," Savory said. "The rancher had lost a half inch of soil because of erosion between the plants."

Savory's verdict? That the prizewinning land was up to 90 percent desertified. Alarmed, the rancher contracted with Savory to help him change the management practices on his land.

Over the years, Savory developed an approach for healing the land that is now known as holistic planned grazing. It departs from all other land management strategies in that the process begins by asking farmers and ranchers to describe the lives they want to lead based on their deepest cultural, spiritual, and material values and to identify ways of working on the land that would support these lives for thousands of years. (In case this sounds excruciatingly earnest, a holistic management trainer who was visiting the center showed me photos of Masai warriors wearing not much more than their beads undergoing his training; it made as much sense to them as it makes to hip urban permaculturists where I now live in Portlandia.) Savory first encountered this kind of holistic framework when he studied the work of Jan Smuts, a South African lawyer, botanist, soldier, and two-time prime minister of South Africa. Later praised by Albert Einstein as having ideated the other of the two concepts—including his own theory of relativity—that would be important to the future of humanity, Smuts believed that nature functions in wholes and patterns, that it is a complicated system that humans erroneously approach as a complex machine. When we interact with nature mechanistically and assume we need only remove or change one cog to fix a problem, we are bound to incur unintended consequences that are often worse than the problem.

An example: Savory likes to point out that many millions of dollars are spent trying to eradicate noxious weeds with sprays and culling. But people engaged in these battles against nature don't understand that they're battling a symptom of the loss of biodiversity in the landscape. "Leaders in Montana spent over $50 million trying to kill knapweed," Savory told a reporter from *Range* magazine in 1999.

"They may as well proclaim it the state flower because there are now more than ever."

Our ignorance about the holistic nature of our environment is best expressed, Savory said in remarks at the 2012 annual congress of the Grassland Society of Southern Africa, by the way we approach our most pressing environmental problems. "The three great issues of the day are biodiversity loss, desertification, and climate change," he said. "Each is being addressed separately by different institutions and even within such institutions—universities, environmental organizations, governments and international agencies, and in separate international conferences. Yet they are one and the same inseparable issue."

Savory's holistic management process takes many variables into account. To manage an ecosystem (Savory says the process can be used to manage anything, from a family to a small business), decision makers look at the array of tools people have traditionally used to manage the land. Savory lists three that are typically used to manage large landscapes: fire, technology (from plowing to spraying chemicals), and rest (from roping off parklands for decades to crop rotation). Instead of these tools, which can never heal brittle lands, he proposes grazing and animal impact—careful grazing in which domestic animals are moved through a landscape as a proxy for the ancient herds that helped build the grasslands in the first place.

I heard much of Savory's story during long interviews around a fire pit in the cluster of thatch and stone buildings where he and Butterfield live during their trips to Zimbabwe. Savory kept a pair of binoculars by his side and peered occasionally at animals moving through the brush. Monkeys peered back at us, a couple of baboons loped by and gawked, and elephants had come calling one night and the ground was littered with branches that they'd pulled from the trees. Savory told me that they have to be careful of lions if they get up in the middle of the night. A warthog came into his hut while he was napping one day and gored him, but that wasn't a wild animal attack—until that day, the warthog had been a camp pet. Savory doesn't do anything to remove the dangerous animals unless one becomes a prolonged nuisance. He respects the role of predators in keeping the wild herds wild so that they bunch and travel and heal the land. Sedentary elephants are as bad as sedentary cows. In fact, Savory is an enthusiastic predator himself. After my last interview, he shared one of his favorite game dishes with me: smoked elephant trunk from his freezer. Perhaps I would have enjoyed it if it had been thoroughly thawed, or maybe my taste buds were held too firmly in the grip of my fondness for Dojiwe, the young orphan elephant that was the new camp pet. Her name means "lost and found."

Savory has many speaking engagements around the world, but many people visit him in Zimbabwe—scientists, cattlemen, politicians, filmmakers, journalists, environmentalists, and more—and many sit right where I was, near the fire pit, with the baboons and other animals looking on. When I visited, the center staff was still excited that Savory had won the 2010 Buckminster Fuller Challenge, which targets solutions to big global problems. They were stoked at the possibility of securing an even bigger honor: The center is a finalist for billionaire Sir Richard Branson's $25 million award for the best plan for removing carbon dioxide from the air. Shortly after I returned to the States, a video of Prince Charles expressing his interest in Savory's work was circulating.

"I believe this is the first real breakthrough in thousands of years," Savory told me. "When you get a breakthrough of this magnitude, you never make it on your own. I was building on my own mistakes, building on the mistakes of others. We can learn from our mistakes if we're open."

I imagine that comments like that, as well as Savory's repeated statements about the uselessness of conventional rangeland science, continue to rankle many in mainstream and academic science. Others are more accepting, but still feel that Savory's methods have not been adequately proved by well-controlled studies. In November 2012, a group called Managing Change Northwest invited Savory to speak to the Washington Cattlemen's Association, the Tilth Producers of Washington, and the Seattle public. Chad Kruger, director of Washington State University's Center for Sustaining Agriculture and Natural Resources (CSANR), helped organize the event, then posted his reflections about Savory online. He noted that some critics have concluded that the science supporting Savory's approach "is either anecdotal or statistically inconclusive because the experiments were poorly designed (they didn't isolate single variables for analysis). . . . While I don't think that critiques are necessarily a definite dismissal of the legitimacy of HM [Savory's holistic management system], I do think they raise an extremely valid question . . . 'Where's the data?'"

Savory responded online, "Your struggle to understand is natural in a paradigm shifting situation. . . . It took me many years because I too was blinded by my reductionist university education." He referred Kruger to a number of reports and studies, only one of which was the kind that Kruger had been looking for.

It is hard to test Savory's methods using conventional scientific studies, which in this kind of situation might compare two identical plots of land with one changed variable—say, the amount of water or the number of animals on the land—and see how the plots diverged on a list of criteria a few years later. But if you isolate one variable, then you're treating the environment as a machine rather than a complex system—and once you do that, you're no longer talking about holistic management. Even Keith Weber, the author of the NASA–funded study that Kruger liked, says that the holistic management system that he and his colleague tested against two other approaches—total rest and a rigid rest/ grazing system—wasn't really true to Savory's approach.

"Truly, if we were doing holistic planned grazing we would have to monitor and change as we go along," Weber told me. "With Savory's approach, you have to monitor how the landscape is reacting to your management and decide on an annual basis how you're going to fix any problems you accidently created. But when you do a scientific experiment, you can't change what you're doing one year into the experiment. You have to do the same thing every year so that you can compare practices. If you make a change, your scientific analysis is thrown out the window."

Weber says that another problem with testing Savory's approach is that healing the land with cattle takes 5 years or longer, but most scientific experiments are funded for only 3 years. Sometimes scientists can get funding for another 3-year study to piggyback on top of the first one, but that's hard to do.

But even though the Savory approach hasn't been proved to a skeptical scientific mainstream, many ranchers have been impressed enough by Savory's ideas to attend his training sessions and turn their own lands over to holistic management. Even without the blessing of mainstream science, the number of these practitioners keeps growing. Savory and Butterfield know of 10,000 ranchers doing holistic management, although there may certainly be more.

The whole point of the Zimbabwe center, as well as two learning sites in the United States and a growing number of sites around the world, is to show the method at work. Savory bought the center's land 40 years ago from a farmer who ran 100 cattle over his 6,500 acres (now expanded to 8,650 acres plus an additional 2,500 that the center manages but does not own), just 25 miles from his original home, which is now a national park. When Savory decided to turn this property into a learning site in 2002, he began increasing the number of cattle. When I visited, the center had 500 cows, including their own along with cows that belong to its herders and to nearby villages. Savory planned to double that number in 2014 in an attempt to replicate the large herds that used to keep the rangeland healthy. "We don't have enough," he told me as he frowned at an area with shoulder-high yellow grass. "I want this trampled to the ground."

Savory had taken me and two goat ranchers from Australia on a tour of the property, along dirt roads so rutted I was afraid I'd break a tooth as we rattled into the bush. Occasionally, he stopped the SUV and we'd get out to observe the land. We stopped once to look at bare ground with such a hard crust on the top that it made a ringing sound when hit with a stick. Savory told us that 50 percent of the land on this property used to be like that. He stopped to show us areas that he'd carefully photographed before they were treated—he calls planned grazing a "treatment"—and invited us to compare the photographs with what lay before us. Using fixed points like forked or fallen trees, we saw that the same patch was now denser with yellowed grass. "It used to be so bare here that you could shoot a guinea fowl at 100 yards," Savory said. "There was nowhere for them to hide. Now they can hide at about five yards."

I asked if he could feel the difference in the soil itself with his bare feet. He nodded. "When the land is bare, it gets awfully hot," he said. "That's not good for plant life, but people with shoes aren't aware of how damned hot the soil gets."

We drove through miles of yellow grass and black trees, stark yellow hills wrinkling up around us. We passed gray patches of ash that his staff throws out so the birds can dust the lice off their feathers. We passed Meg's Spot, the place where back in the 1970s he left his 6-year old daughter with a rifle to guard some tourists while he trudged off to find someone to repair his broken-down vehicle. We passed a tumbled-down rock shack near the Dimbangombe River where he had left some visitors to watch the elephants bathing and a group of lions climbed on the roof to share the spectacle. We visited the kraal, a large circular enclosure made of white plastic where the livestock spend the night, moved weekly for concentrated doses of animal impact across the land. It was empty when we reached it, a bit odd and Christo-like out there in the wilderness. One of the herders came out from his nearby tent with a few of the Jack Russells that yap a warning at night when lions, elephants, and hyenas approach. They don't accompany the herd anymore, as one of their numbers was eaten and another bitten by a puff adder. After some driving off-road, we finally came upon the herd itself, which is taken from the kraal each morning to graze a certain part of the range for about 3 days—the whole range is mapped out and divided according to the annual grazing plan.

I had to keep pinching myself and saying, "Africa!" even though the yellow land reminded me of the dry winter landscape around my childhood home in northern California. There were so many reasons to be thrilled about being there. I rode the baby elephant, Dojiwe. I visited Victoria Falls. I woke up one night to the sounds of hyenas howling outside.

But the most thrilling moment was when Savory took us to the river that had been dry when he first bought the property. It had been dry for as long as anyone in the nearby villages could remember. It was dry in satellite images from years back. There, springs trickled from the dry winter landscape into the river, now muddied with an abundance of elephant tracks. The thick grass we saw aboveground was a symptom of the deep healing going on in the soil, where microorganisms were building aggregates that both received and held on to water. The land was now a mighty reservoir, and I thought it would not return to desert anytime soon.

CHAPTER 4

LETTING NATURE DO ITS JOB

A fleet of dusty pickups and SUVs bumped across the North Dakota prairie, the summer grasses and wildflowers and wiry forbs susurrant against their undersides. From my seat in Jay Fuhrer's SUV, it sounded as if we were in a boat cresting a river of chop. We finally pulled up next to one of Mike and Becky Small's cornfields, jumped out into the lemony air—the vehicles had crushed a lot of wildflowers called lemon scurfpea—and formed a ragged circle of plaid shirts. At the center stood Fuhrer, the USDA Natural Resources Conservation Service conservationist for Burleigh County, the county seat of which is Bismarck. The sun was blinding, and I was the only foolishly hatless person in the group. I hunkered in a tall guy's shadow, mindful of the tobacco juice he kept spitting to the side.

Fuhrer is a compact man with some silver in his hair and a wryly self-deprecating habit of referring to himself as "the old German." On that July morning, he could have doubled as the genial host of a cooking show. He bent down and carved up a brownie-size chunk of soil from the Smalls's cornfield, broke it apart and waved it in front of his nose as if savoring the complexity of its ingredients. He passed dark chunks around for everyone else to sniff and appreciate. Then he wrenched up a cornstalk and shook it until most of the soil fell away from its roots. Even with all his shaking, the tangled strands were still coated in a thick, dark layer of sticky soil. They looked like dreadlocks.

"Why doesn't the soil fall all the way off?" Fuhrer asked the crowd, touching the roots. "The glues in the soil hold it there. There are aggregates being formed right now."

He broke off one of the larger roots and asked someone to pour their bottled water over it. When it finally washed clean, he sliced it into pieces and passed them around the crowd, like hors d'oeuvres. I popped a piece in my mouth, and it was—perhaps unsurprisingly—cornlike, sweet and crisp and cool. "Can you taste those sugars?" Fuhrer asked. "Those are the soil exudates! That's what the plants use to attract the biology."

Just as people from all over the world visit Allan Savory's holistic grazing model on the grasslands of Zimbabwe, people from around the United States and beyond visit Burleigh County to see how soil health can be built in land that is actively cropped. A renegade band of 40 farmers and ranchers there—I'm not sure what they should be called, as most of them both plant crops and raise meat animals—with enthusiastic backup from Fuhrer and USDA scientist Kristine Nichols have done what nearly everyone believes is impossible: They are building healthy, carbon-rich soil and healing their landscapes while increasing yields and making greater profits. And, as Mike Small told me and the crowd of farmers and Natural Resources Conservation Service employees from Missouri who toured that week, they also enjoy more time with their families.

By its very nature, conventional cropping is a far greater assault on the environment than herding animals. Plowing rips apart the crucial underground networks of mycorrhizal fungi and shatters the soil aggregates that hold water and gases in the soil. After these aggregates break down, the soil particles pack tightly against one another—this is called soil compaction—and the land can't capture and hold the water from either irrigation or rain. In fact, a recent study showed that nearly half of the rise in sea level comes from water that runs off agricultural lands. Want to know where the quickly draining Ogallala Aquifer is going? Lots of it winds up in the ocean. About 70 percent of America's freshwater usage goes to agriculture, but compacted soil means that much of it can't penetrate. Tillage equipment is redesigned periodically to strike deeper and deeper to break up this compacted soil, but this only creates a new and still deeper layer of compaction.

In the process of preparing a field for sowing, conventional farmers also remove all the vegetation so that they can offer a blank slate to the one crop they want to grow and sell, whether it be corn (planted on 24 percent of America's 406 million acres of cropland) or wheat (14 percent) or soybeans (19 percent). Weeds, other plants, and even residue from a previous year's crop are removed, often in the fall so that the farmer can plant more quickly the following spring. This leaves the soil bare and exposed for up to 7 months. The process wasn't designed to starve the soil microorganisms, of course, but that's what it does, since there are no live roots in the soil to feed them exudates and no dead plant materials lying around for them to gnaw on. When I drove from Cleveland to Portland in the fall of 2012, I sped past what seemed like thousands of these naked brown acres. Sometimes I passed the culprit: a tractor pulling a huge disc, pluming off so much dust that it was hard to see the highway. It was almost as if I was downwind from a fire.

Even most organic farmers conduct this annual ruination of the soil, especially the huge industrial concerns that produce most of the organic products in our grocery stores. They can't call themselves organic if they use chemical herbicides to get rid of weeds, so they till them away.

Clearing and tilling farmland has been going on for millennia—some of the world's poorest soils and communities of people have been created this way—but today's machinery allows it to happen at a far more massive and accelerated scale. Farmers drop seeds into this degraded soil come spring, but have little hope of growing a crop in ground where all the natural processes have been devastated. And not just by tilling and clearing. Without a healthy community of soil microorganisms to provide nutrients, something has to be added. Organic farmers rely on manure, compost, or natural fertilizers to restore some of the lost nutrients, but most conventional farmers—and about 99 percent of our food comes from them—have been subjecting the land to a harsh chemical bath for years. That's what nearly every expert they've ever encountered has told them they must do to survive.

As entrenched as chemical farming seems, it's only been around for about 50 years. As with so many innovations, the "process of taking atoms from the atmosphere and combining them into molecules useful to living beings," as Michael Pollan writes in *The Omnivore's Dilemma,* is connected with the exigencies of war. Nitrates are needed to make bombs, and a German-Jewish scientist named Fritz Haber figured out how to make synthetic nitrates for the bombs and poison gases that were used during World War I. (He also developed the poison gases later used in concentration camps during World War II, although the Nazis had forced him to leave the country by then.) Weirdly, Haber's work to pull nitrogen from the air was originally undertaken to create chemical fertilizers and boost agricultural productivity. His invention "liberated" agriculture from biological processes and allowing farmers to raise crops without needing much knowledge of natural systems. "Once chemical agriculture came along, you didn't need any skill, you didn't even have to know how to be a farmer," says Abe Collins, a farmer and soil visionary from Vermont. "You could just throw that stuff out there, even on really degraded land, and get a crop."

Picky consumers like me look for the organic label in a grocery store or, better yet, a farmers' market, because we have an intuitive feeling that foul-smelling chemical fertilizers couldn't possibly make healthy food—we feel that nature's way has to be better, although we don't really know why. But the new scientific understanding of what's happening in the soil validates this intuition. Most chemical fertilizers are a mixture of the three minerals that agricultural scientists long ago determined are essential for plant growth: nitrogen, potassium, and phosphorus. But as microbiologist Elaine Ingham points out, as the tools of science get better, scientists pinpoint more and more nutrients in foods that are important for our health. Those nutrients aren't going to reach the plants through the application of these chemical fertilizers, because they're not in the mix. In fact, the full panoply of necessary nutrients might *never* be in the mix, because the interactions between the plants and the soil microorganisms—nature's way of providing plants with the minerals they need—are so very complicated and hard to replicate.

Even after tilling, soil microorganisms will still be in the soil, but they aren't likely to provide these varied nutrients to the plants once the chemical fertilizers are applied. Simply put, these applications interfere with one of nature's great partnerships. By the terms of this partnership, plants are supposed to distribute carbon sugars through their roots to the microorganisms in exchange for nutrients. Fertilizer disrupts this pay-as-you go system. Plants get lazy.

"When we add fertilizer, we're putting nutrients right next to the plant roots and the plant doesn't have to give up any carbon to get them," says USDA microbiologist

Kristine Nichols. "Therefore, the soil organisms can't get enough food."

Without their carbon meal, the mycorrhizal fungi can't grow and stretch their strands of carbon through the soil. They and the other soil microorganisms can't produce the glues that fix carbon in the soil and build the aggregates that hold water. They go dormant and, given enough stress, can die. At that point, the soil is so depleted of life and structure that a farmer *can't* get a decent crop without chemical fertilizers, at least not for a couple of years. "Then we get into a system where we're adding more and more fertilizer to try to maintain or increase our yields," Nichols says. "We'll see symptoms that look like fertilizer deficiency when we don't add enough, because there isn't that beneficial activity with the organisms."

Conventional agriculture uses about 32 billion pounds of chemical fertilizers every year, but the fertilizers are remarkably inefficient. Much of the phosphorus in chemical fertilizers quickly binds to minerals in the soil and becomes unavailable to the plants. Soil microorganisms have enzymes to make the phosphorus available, but they're often dormant or dead under a chemical regimen. The problems associated with nitrogen uptake are even worse. Without healthy soil biology to convert the nitrogen into a plant-palatable form, up to 50 percent of the nitrogen is lost, washed away with the rain or irrigation water into the groundwater or streams. There, it enriches the waters and causes algal growth, which sucks oxygen out of the water and creates dead zones. The Gulf of Mexico has one of the largest dead zones in the world—about 6,000 square miles near the mouth of the Mississippi—caused by fertilizer run-off. The one benefit of the drought of 2012 was that the Gulf dead zone shrank because not as much nitrogen-enriched water rolled down the river.

Typically, conventional farmers react to the poor absorption of their chemicals by adding more. They'll add 100 pounds of nitrogen just to get 50 in the soil.

Most farmers are discomfited by the downstream effects of tillage and fertilizer use, but everyone—from their agriculture school professors to their county extension agent—has been telling them for years that *this* is the way to build a successful business. Now the rising price of chemical fertilizers, which are fuel intensive to make and apply, has many farmers and the people who work with them looking for a better way. Thus the allure of Burleigh County: These farmers have gone back to working more closely with nature because their crops are as good as or better than—usually better than—when they went the chemical route, and they save thousands of dollars not applying chemicals.

Our final stop on Mike and Becky Small's farm was at a field so densely planted that it wasn't possible to even see the soil. Just before we climbed back into our vehicles to head for the next farm on the tour, a young man in a blue plaid shirt and camo cap raised his hand. "What are your inputs here?" he asked Small, who just grinned.

"Inputs?" Fuhrer frowned as if he didn't understand.

The young guy fell into the trap. "Yeah, inputs. Your fertilizer."

"None," Small tells him.

"None at all?"

"That's right."

The young man took this in, then sighed. "Will you come down to Missouri and talk to my dad?"

How did the Smalls manage to raise up this jungle of biomass without fertilizer or, for that matter, irrigation, in a part of the country that averages 15 inches of precipitation per year? The answer began to form nearly 15 years earlier, on Gabe Brown's farm outside Bismarck, long before the Smalls thought it was possible to depart from conventional farming. There, nature had a showdown with Gabe Brown on the 5,400-acre property he farms with his wife, Shelly, and son, Paul. The Browns completely changed the way they farm, and their discoveries are rippling through the county and beyond.

I had actually planned to visit Gabe Brown even before I found out about Fuhrer's soil tour. Two years earlier, I attended a conference in New Mexico hosted by the Quivira Coalition, a group that convenes progressive ranchers and environmentalists to talk about how good agriculture can solve many of the environmental problems both worry about. The theme of the conference was "The Carbon Ranch: Using Food and Stewardship to Build Soil and Fight Climate Change"—totally in line with my interests. Still, I waited to sign up until the last minute, so I didn't get a room at the hotel where the conference was held. This may be why I'll never give up procrastination. I had to stay in an off-site motel and catch a shuttle every morning to the conference, and, because of that, wound up sitting next to Eliav Bitan several mornings in a row. At that time, he was the agriculture advisor to the National Wildlife Federation's Climate and Energy Program. The NWF had declared that the greatest threat to wildlife was global climate change, and Bitan had the enviable task of going around the country to study the work of progressive farmers and cull "future-friendly" farming practices for a report. We stayed in touch and finally met again in North Dakota so he could introduce me to Gabe Brown.

I flew out in July of 2012, at the height of the droughtiest summer in 50 years, wondering as my plane circled the Bismarck airport about the rigid sentinels of trees marking up the countryside. From the air, it looked as if someone had thrown down a handful of pick-up sticks. When Bitan and I arrived at Brown's farm, I found out that these were massive tree breaks, one of the conservation practices instituted in the wake of the Dust Bowl years of the Great Depression. In the years leading up to the Dust Bowl, farmers had become reckless "sodbusters," using the biggest machines they could find to tear up the native prairie and make their fortunes planting wheat. They did nothing to care for the soil; instead, they kept raking it open for more and more crops. These bad farming practices ultimately collided with drought. Instead of rain, they got powerful winds that whipped the bare soil into enormous dust storms that blocked out the sun for days and sickened people and animals. North Dakota had been hit hard by this man-made "natural" disaster. On Jay Fuhrer's soil tour, a visiting scientist from Missouri speculated that Dust Bowl winds had resurrected boulders that were now lying around in the fields from deep under the soil, where glaciers had deposited them millions of years before.

But everything was green at the Brown farm, even with only 8 inches of rain since the beginning of the year; crazy green, green in a kind of horticultural chaos that didn't even slightly suggest the neat geometry we associate with farming. I think I had an initial dismayed moment that it wasn't pretty. Where were the tidy rows of crops? Where were the smooth green fields decoratively dotted with black-and-white cows? Instead, there were uncropped fields that were thick with vegetation, an Eraserhead profusion of plants jammed together and—except for the occasional sunflower nodding its bright face over the fray—unidentifiable from a distance. Even the cornfield was messy, with both living and dead plants crowded between the rows of corn. And the livestock? They were a multicolored gang of cows and calves bunched together on a small piece of Gabe's 5,400 acres, cordoned off with mobile electric fencing into an inelegant configuration. The only tidy spot was the yard around the Brown's house, which Gabe's wife, Shelly, was circling on a riding mower.

This was one of those moments when I realized I had to retrain my sense of beauty when it came to farming. As

Marlyn Richter, one of the other farmers we visited on the soil tour said, "Our dad thought we were nuts when we started doing this. He wanted the place to be pretty, with neat brown rows. But now, the uglier it looks, the better we feel about it."

Gabe Brown is a broad-shouldered, sturdy man with a Northern Plains drawl and a quick laugh, who always looks as if he's peering into the distance, figuring out his next experiment. He wasn't born to the art and science of farming—he grew up a Bismarck city boy—but loved the outdoors and took a vocational agriculture class in ninth grade. He became hooked on farming and spent his summers working at a dairy farm 50 miles away.

Shelly's father was a farmer, and he sometimes hired Brown to clear his fields of rocks. Gabe and Shelly didn't date in high school; she knew she didn't want to marry a farmer and only wanted friendship with Future Farmers of America types like Gabe. But by the time they finished 2 years at Bismarck State College, they were wed. Gabe went on to get a 4-year degree from North Dakota State in Fargo, where he majored in animal science and agricultural economics. Even though he was passionate about these subjects, he winced at the way they were taught. "It didn't seem like the puzzle was fitting together," he told me, easing back in his camo-covered lounger during a long interview in his basement. "Each enterprise on the farm was separate and we weren't looking at the bigger picture. I was frustrated with that, but at the time I didn't know why."

Gabe and Shelly moved to her parents' farm to help

them manage it, and they purchased a part of the farm in 1991. At that point, Brown was a conventional farmer, not a revolutionary one. He tilled, he applied fertilizer, he sprayed pesticides, herbicides, and fungicides, and he hung fly-killing ear tags on his cows.

By 1993, though, tilling seemed wrongheaded to him. "It didn't make any sense that we dug up the soil and then complained later that it was too dry," he explained. "And since I didn't grow up on a farm, it wasn't ingrained in me to do exactly what Dad and Grandpa did." He talked to a no-till-farming friend—there weren't many in the county back then—and the friend advised him that if he was thinking of going no-till, he had to sell his plows. Otherwise, he'd always be tempted to go back to tilling.

So Brown sold his tilling equipment. He bought a shiny yellow-and-green John Deere 750 no-till drill that makes a tiny slit in the soil, drops a seed inside, then quickly seals the gap. He started to depart from the ag school regimen in other ways that made sense to him, too, such as sowing legumes in his pastures to fix nitrogen in the soil. He was disappointed with the amount of forage growing in 200 acres of "tame" grass—pasture seeded with store-bought grass as opposed to native prairie grasses—and sought advice for bumping up production. The standard advice from agricultural extension educators was to add fertilizer, but he couldn't afford more fertilizer. He consulted with Jay Fuhrer and they decided to seed 12 pastures with different legumes—alfalfa, cicer milkvetch, bird's-foot trefoil, and

clover, as well as eight others—to see which improved production the most. "We got tremendous production," he
said. "That showed me the synergy between grasses and
legumes both in cropland and pasture. We had a lot of people coming to visit these pastures because they were way
more productive than monoculture tame-grass pastures."

And then he had his epic, crushing, biblical-level bout
with nature. For 4 years in a row, weather anomalies—either
hail or late frost or extreme drought—wiped out most of his
crops. For 4 years, he couldn't take much of anything to
market. To make do, he sowed his ruined fields with summer crops just to have some feed for his cattle. He tried
sowing corn in 1997, but the plants withered in a drought.
Instead of clearing the field mechanically with a tractor and
disc, he sent in his cows to eat the stalks. He also sowed
cover crops in the fall—these are plants grown in the market-
crop off-season to protect the soil from wind and rain
erosion—hoping they might improve moisture. He noticed
that his soil health seemed to be improving, but really, survival and hanging on to the farm were his only concerns.

"That fourth year, we lost 80 percent of our crop to
hail," Brown recalled. "It's awfully tough to make bank payments when you have no income. That was a tough time,
but it was the best thing that could have happened because
I never would be where I am today without those 4 years.
We were forced to change."

By this time, Brown couldn't afford to buy fertilizer—
he was broke—but he realized that his departures from

conventional agricultural practices had been improving his soil so much that he might not even need it. So he kept looking for other ways to farm and ranch successfully without expensive chemicals. In 1998, he took a course in Savory's holistic management methods. He learned that set stocking or season-long grazing—those are terms for what most ranchers do when they turn out their animals to wander one large pasture for the entire year—encourages the cows to eat their preferred grasses past the point of recovery and allows weeds to take over. He began the process of turning his pastureland into more than 100 smaller paddocks and rotating his cattle through them every few days, allowing the grasses plenty of time to regenerate.

Right around that time, Brown became a supervisor with the Burleigh County Soil Conservation District, and he and Jay Fuhrer became a convention-busting team, eager to test any new ideas that sounded promising. Brown expanded his cover-crop mixes to include a few more varieties, and word started to spread about his healthy soil.

Then Kristine Nichols, at the time a new soil microbiologist with the USDA's Agricultural Resource Service in nearby Mandan, came to visit. Impressed by what she saw going on with Brown's soil, she urged him to cut back on the amount of chemical fertilizers he was using (now that his farm was more successful, he'd gone back to using fertilizer). He recalls her saying, "Gabe, you've got to reduce and eventually eliminate your commercial fertility so your soil biology will function as it should. That's the only way your system can become sustainable."

Brown embarked upon an extended trial to test her suggestion, using fertilizer on half of each individual field and leaving it off on the other half. After 4 years, the halves without fertilizer consistently outperformed the other halves, so it was easy to decide to ditch the cost and bother of applying fertilizer. Clearly, all the other steps he'd taken—not tilling, the cover crops, the occasional foray by cows to eat crop residue—had changed the soil so that his crops didn't need fertilizer. Soil health became an obsession for Brown at that point, with Fuhrer and Nichols his ever-eager partners.

In 2006, Brown and Fuhrer attended a No-Till on the Plains conference in Kansas. A Brazilian crop consultant named Ademir Calegari took the dais to talk about the cover-crop "cocktails" farmers were using to build soil health in South America. Cover crops themselves have been used by farmers for thousands of years to curb erosion and runoff; the leaves catch raindrops and allow them to drip on the soil slowly, preventing them from hammering soil particles apart and giving the water more time to percolate down. Despite these benefits, cover crops are not widely used in the United States. In fact, a National Wildlife Federation study showed that in 2011, only a maximum of 4.3 million acres of the Mississippi River Basin's 277 million acres was planted in cover crops.

Gabe Brown had noticed that water runoff from his cropland dropped dramatically after he started using his simple two- and three-species cover-crop mixes, and as he learned more about the science of soil health, he suspected that the cover crops were doing more than just reducing erosion and runoff. The cover crops ensured that his soil biology was being fed throughout the year, not just while he was growing his commodity crops of wheat and corn. With these underground engineers building soil aggregates and organic matter year-round, his land had become spongelike, soaking in and holding water so that his crops thrived even during dry weather.

The cocktails that Ademir Calegari suggested bumped the cover-crop concept to a new level. One of the most unnatural practices of conventional farming is its creation of vast landscapes in which only one plant is grown, usually corn or wheat or soybeans. But such "monocultures" never occur in nature. A square foot of the native prairie near Gabe Brown's farm has up to 140 plant species. An acre of land alongside any American highway probably has more plant species than all of Iowa's cropped lands put together. A lush mixture of plants aboveground means that there is a correspondingly lush community of microorganisms underground, as the different plants offer different root exudates and attract an array of different microorganisms, making the soil overall more resilient. Nature is always trying to restore balance to landscapes that humans degrade. When we create bare land, nature sends a battalion of weeds to colonize and cover the soil. When we establish a monoculture, nature sends in pathogens to weaken and even kill that crop, allowing other species to fill the void.

"If you grow a monocrop for an extended period of

time, you're promoting pathogens that are specific to that crop," says John Klironomos, a biologist at the University of British Columbia. "This stimulates diversity. Nature won't let any one species dominate a particular area, but pathogens won't find a host in a very diverse agricultural system." Ironic that we immediately think that crops are sick when they're afflicted with a pathogen, when this is nature's way of making the landscape healthy again.

Gabe and Jay left the No-Till on the Plains conference eager to put Calegari's concept into trials. They went to the grain elevator, which not only receives and pays for commodity crops but also sells seed. They flummoxed the folks there by requesting 1,000 pounds of turnip seed. "*How many seed packets do you want?*" was the astonished reply. But finally, Brown and Fuhrer gathered up enough bulk seed to plant test fields on conservation district land, some as monocultures and some as cocktails. Soil health seemed to improve with the cocktails. The drama came during the dry summer of 2006, when the county got only an inch of rain between the time that they sowed the trial fields in May and clipped and weighed the growth in late July, some 70 days later. Most of the monoculture crops died, but the cover-crop cocktail fields had grown like mad. "That showed us that there was a tremendous beneficial influence from growing these crops together," Gabe said. He began planting cover-crop cocktails himself.

The next arc of his learning curve came in 2008, when he spoke at a grazing conference in Manitoba. A rancher named Neil Dennis from Saskatchewan approached him after his talk and told him about his experiment with another innovative practice called mob grazing. Brown typically ran 50,000 pounds of cattle on an acre—about 40 cows—but Dennis was running an astounding million pounds of cattle per acre. Brown sat up until 3:00 in the morning looking at photos on Dennis's computer. The following June, he visited Dennis's operation and was stunned by the health of his soil.

The mob grazing concept builds upon the insights of André Voisin and Allan Savory, namely that even very large numbers of animals are good for the land if one controls the amount of time they spend there. With carefully monitored grazing times that gave the plants time to recover, Dennis found that the beneficial impacts of herbivores on the land were multiplied by having *lots* of them there: lots of hooves breaking up the hard surface of the soil, lots of grasses trampled into the ground, lots of grasses being tugged and bitten and causing the plants to pulse carbon sugars into the soil, and lots of nutritious dung, urine, and hair spread around for the insects and microorganisms to break down.

Now, the Browns have a winning trifecta of soil-health practices: no-till planting combined with cover-crop cocktails, followed by mob grazing. They currently turn nearly three-quarters of a million pounds of cattle onto their pastures and also let them storm the cover-crop-cocktail fields. Not that there haven't been other innovations, as well— Brown says he wants to make sure he and Paul fail at one new thing every year, maybe more, just to make sure they're stretching. For instance, most ranchers arrange their cows' breeding cycle so that they give birth in late winter, inside corrals. The Browns decided they wanted their cows to give birth in May, out on the fresh grass and in the sunshine. "Out there, the cow is consuming a healthy fresh diet instead of hay or something that took fossil fuels to get in front of her," Brown said. "And it's a clean environment for the newborn, no different from that of the bison or antelope or deer." This innovation wasn't a failure!

There are now chickens on the farm as well as sheep, adding to the diversity of impact on and inputs into the land. Basically, that's what the Browns are always looking for: ways to stack enterprises on the farm so that they make more money on the same acreage and keep building soil health. When Gabe Brown first met Kris Nichols, she told him that the herd under the soil was just as large as the herd on top of the soil, but far more important for the long-term success of his operation. "You have to think about managing the organisms in the soil the way you think about managing for any healthy herd," she told me when I visited her at the Mandan research center. "They need constant and high-nutritive-value food sources. They need a good habitat. They need protection from diseases and predators. When Gabe manages his cattle, he's utilizing them as a tool to manage the soil for that other herd."

When Gabe Brown takes visitors around, they admire the cows munching amiably on acres of green. But he's thinking about that much larger herd down in the soil, steadily lapping up the carbon sugars and turning them into black gold.

By the time I visited him in the summer of 2012, Brown was famous—between his speaking gigs and visits from scientists and farming consultants, it was like trying to get on a rock star's schedule. His son, Paul, had made T-shirts to mark this transformation, navy blues with Brown's face and GABE MANIA! on the front in white and WORLD TOUR 2012 on the back, along with a list of all the places where he had spoken. Gabe modestly declined to show me one. He was producing 127 bushels of corn per acre, 27 more bushels than the county average, without fertilizer, pesticides, or fungicides, and with just a small amount of herbicide. He was spending between $1.00 to $1.25 to produce that bushel, whereas the county's average cost per bushel was between $3.00 to $3.50. The corn and wheat were sold to standard commodity markets, and the cattle to specialty buyers of grass-fed beef. The soil organic matter in his cropped lands has gone from 1.7 to 5.3 percent, which he thinks is not nearly high enough—he'd like to get the croplands as high as his pastures, which are at 7.3 percent. His son, Paul, is determined to get both up to 12 percent. While other farmers continued to think that they needed to buy more land and become larger to be successful—that's been the assumption for years—Gabe Brown decided to become smaller. He was making enough money that he could sacrifice acreage—he discontinued leases on 640 extra acres—allowing him to spend more effort fine-tuning his management.

Gabe Brown was scowling at the sky when Eliav Bitan and I arrived. A silvery crop-dusting plane streamed a contrail of pesticide over a distant field of sunflowers. Brown was as dismayed as if it were the Wicked Witch of the West from the Oz fantasy spelling, "Surrender Dorothy" in the sky. "That just makes me sick," he said. "I don't want that stuff drifting over on my crops."

Then we climbed into his black four-door truck, his border collie Pistol leaping into the back just as we pulled away from the house, and raised our own contrail of dust. We headed off to see a 60-acre field that Paul had sowed with cover-crop cocktails, half with 19 species and half with 26—another trial to educate themselves about soil health. We drove past the odd nomenclature of North Dakota country roads—they had names like First Avenue South and 102 Avenue Southeast to help firemen and police map their way to far-flung homes, but the names looked odd on intersections with no buildings in sight. We drove past fields that were green and fields that had already had their hay harvested and spun into big golden cylinders.

Paul's field was one of the green ones. Sunflowers raised their fist-size buds high. So many shorter plants of varying heights crowded around them that it looked as if the sunflowers could have stood upright, even without roots. When I bent in to look closer, I saw tiny flowers—white and yellow and blue—dangling in the green. I spied pinkish disks of turnip and radish in the soil below. "We're going for a forest canopy structure here," Brown said, more for my benefit than Bitan's—he'd heard all this before. "Like a rainforest, with high, medium, and low layers to the canopy. And we want different leaf types, too, different shapes. They're my solar collectors. No matter what angle the sunlight is, we're collecting the maximum amount of energy."

He reached out to touch a dainty blue flower. It was flax, which they plant because the cattle don't like to eat it. In fact, they only leave the cows in the field long enough to eat 25 percent of the plant material (and yes, they eat the turnips but just nibble at the radishes). The rest is stomped to the ground for the other herd. Brown reached for a sprig of buckwheat, which he and Paul plant because it scavenges phosphorous from the soil and makes it available to other plants. He pointed out the plants that had 8-foot roots and those with shallow ones. Just as they want the plants to fill up the space above the soil, they want the roots to occupy all areas of the soil profile below. "That's biological tillage," Brown said. "That's nature's way of getting rid of soil compaction."

Fine-limbed gray spiders climbed up his arms as he spoke. Behind him, a scrim of white butterflies drifted through the field. When I remarked upon the abundance of insects, he nodded. "We had an entomologist out here the other day, and he was just amazed. He was like a kid in a candy store."

Brown's farm *is* like a candy store to scientists who are interested in studying natural systems in agriculture (or who aren't chemical-spray-loving "nozzle-heads," as some people call the researchers in conventional agriculture). The visiting entomologist was Jonathan Lundgren from the USDA's Agricultural Research Service in Brookings, South Dakota, and he was dazzled by the impact Brown's cover crops were having on his insect community.

"Most farmers think that if they see an insect, they have to kill it," Lundgren told me on the phone. "But there are only 3,000 insect pests in the world that eat our crops and hurt our livestock. For each one of those pests, there are around 3,000 beneficial insects. We don't even have names for most of them yet. Spraying wipes out that whole insect community, including the enemies of those pests. I'd like to see us manage our fields so that we keep those beneficial insects around."

And that's what he's finding in Brown's fields. Brown doesn't know exactly which plants are attracting the predators of the agricultural pests; he just tries to make sure that his cover-crop mixes attract pollinators and that they provide a good habitat for spiders. He assumes that if he maintains plant diversity on the ground, insect diversity will follow. And even though studies haven't shown why this approach is working—Lundgren plans to help with that—Brown knows his approach *is* working. He hasn't used pesticides in 12 years and hasn't needed to.

For instance, Brown doesn't have trouble with the

corn rootworm, a beetle larva that eats corn roots, which Lundgren says is the number one agricultural pest in America. Billions of pesticide dollars are spent every year to combat this pest, but they also harm the rest of the insect community—and they wind up creating super pests when the bad insects develop resistance to the poison. Lundgren has been investigating the stomachs of other insects for evidence of corn rootworm DNA and has identified dozens of predators. "The predator community will be there if we provide the right environment for them," he explained. Gabe Brown's fields are attracting the predators, and Lundgren suspects that something else is going on, too. He believes that the cover-crop cocktails are changing the root structure of the corn plants and inducing the corn rootworm to venture away from the protection of the roots where they are nailed by predators. When I spoke to him in 2013, he had submitted a grant proposal to the USDA to research how, exactly, the diverse plant and insect varieties in Brown's fields are working together to defeat the corn rootworm.

Brown's lively insect population may also be the cause of his dwindling need for herbicides. Insects don't eat just plants and other insects; they're also granivores that eat seeds. Most of the crops that farmers sow have been bred to bear large seeds, which aren't appealing to these small granivores. But weeds have tiny seeds and a vigorous propagation mechanism that shoots out thousands of the seeds—millions in many species—over the course of a growing season. These tiny seeds are a tasty meal to the right insect. Unwittingly, farmers who kill off their insect populations to get rid of one pest may be making it easier for their weeds to dominate.

Gabe Brown doesn't always know exactly why these diverse plant and insect populations have such a positive impact on his business; he just has all the anecdotal evidence he needs to convince himself and many others that they do. But Lundgren enjoys figuring out the science behind the success, especially since no one else has even thought of researching these complex relationships. It's a great partnership. "I've really learned from Gabe," he says. "He's as much driving my research program as I'm helping him, if not more. The stuff he and some of these other progressive farmers are doing is the future of farming."

In fact, one of the things that distinguishes these progressive farmers and ranchers is their eagerness to work with progressive scientists—meaning those whose work is not hog-tied by the vast research budgets of the corporations that make money on fertilizers and other products that work against nature, or by the academic agriculture departments that survive on those budgets. Brown also has a great partnership with Rick Haney, a soil scientist with the USDA's Agricultural Research Service in Temple, Texas.

Haney came from a farming family and always wanted to be a farmer himself, but he couldn't afford to buy land when he left college in the 1980s. So he went into soil science and has farmed his research station's 200-acre test plot for the last 12 years. Unlike the city-raised Brown, he had been schooled by his family to adhere to the rules of conventional farming. "I'd been tilling forever," he told me. "It wasn't until just a couple of years ago that it hit me that I was destroying nature's brilliantly designed underground system to move minerals and nutrients and water."

Haney's real education about the soil came from studying Gabe Brown and other farmers. He soon realized that the way his profession analyzed soil was completely inadequate and outdated. Most of the soil tests that are currently in use were designed between 40 and 60 years ago, and they reflect the mind-set of an earlier age, which viewed soil as a mixture of chemicals rather than a complex, living system. He decided to create a soil test that reflected this natural complexity. "I'm a big fan of nature," he says. "Nature has about 3 billion years of research and development. I try to mimic in the lab what nature does outside."

Soil samples are routinely dried in a kiln so that they all have the same moisture content and are therefore comparable. One of Haney's innovations is that he rewets that dried sample, then measures the amount of carbon dioxide gassed off over 24 hours. When the sample is dry, the microorganisms become dormant. When it's wetted again, they wake up, get back to work, and begin to breathe. That pulse of carbon dioxide reveals the vitality of the microorganism community and the soil's overall fertility. When Haney first tried to publish a paper about this drying and rewetting method, a reviewer at one journal commented, "You can't do that. It's too simple."

Haney also devised a method to get a better bead on soil nutrients. Conventional soil testing uses harsh chemicals to extract minerals from a sample, but Haney felt that these chemicals caused an incorrect measurement because they pull things from the soil that aren't actually available to the plants. So he came up with his own water-based extractant based on the chemistry of the root exudates. His test also looks for all the different forms of various nutrients. For instance, conventional soil tests typically only look for inorganic nitrogen, but Haney had observed that plants were thriving in soils that had only half as much inorganic nitrogen as the experts deemed adequate. Using scientific techniques perfected only in the last decade, Haney's test also measures the organic nitrogen in the soil that the other tests miss.

Haney offers farmers this more accurate depiction of what's actually going on in their soils, coupled with advice about cover-crop cocktails that include combinations of grasses and legumes to get the maximum benefit from the huge pool of nitrogen already in most soils. This allows farmers to cut back on their nitrogen fertilizer—an environmentally sound move and a huge cost cutter for the farmers. "These fertilizers have been used for 50 years as an insurance policy because they were fairly cheap," Haney says. "That's not the case today, so there's a lot more interest in getting by with less."

Plato said that "necessity is the mother of invention" (at least Wikipedia ascribes it to Plato; I was pretty sure it wasn't Frank Zappa!), and farmers and ranchers are facing some stark necessities these days. The rising cost of chemical inputs is certainly one of them; the other is the unpredictable weather patterns caused by climate change. Soil microbiologist Kristine Nichols suspects that the agriculturalists in Burleigh County have been more inventive and receptive to change because they work in such an unforgiving climate: They have a short growing season and frequent drought. But as the rest of the nation found out in the drought of 2012, no one can assume there will be an abundance of rain at the right time. The challenge is to build healthy soils so that every drop of rain goes to good use.

The high point of my visit to Gabe Brown's farm came after we left Paul's field and churned over a gravel road to one of Gabe's cornfields. He had been carrying around a slim 4-foot metal rod. He told me it was a moisture probe, but he had been using it to point out the different plants and insects in Paul's field. Then we walked into the cornfield, which seemed to be at least a foot taller than any of the neighbors', and he nudged it into a bit of bare soil. And then—and then!—he pushed all 4 feet of the rod straight into the ground, all the way up to his knuckles.

"I can't believe that!" I think I dropped my recorder. "Do it again!"

So he walked a few feet away and shoved the rod into the soil again, then pulled it out and held it out to me. "You try it."

My arms aren't nearly as substantial as Brown's. Where his arms bulge with muscle, mine jiggle. Without much expectation of success, I took the rod and pushed it into the ground. I tried it in several places. And each time, I pushed the rod all the way up to my knuckles. I knew what an amazing thing this was, since I've been a backyard gardener ever since I was 25. Even after years of babying my beds with bags of compost, I never had soil like that in Gabe Brown's cornfield. I could hardly stick a fork in my lawn back in Cleveland! But through his management of this harsh landscape, he had created soil that was so rich with microbial life that they had built aggregates going down at least 4 feet. Four feet of carbon-rich soil, stacked with billions of tiny cups to hold water.

Brown shrugged. "I don't worry about drought."

We tried the moisture probe in another of Brown's cornfields as well as in a crazy field where he has 75 species of flowers and vegetables planted for his family and the local food bank—it's a nutritional jungle where you can't take a step without crushing a banquet-size portion of vegetables. The only place the moisture probe wouldn't penetrate was in his neighbor's field, who is a no-till farmer who hasn't tried the cover-crop cocktail and some of Brown's other innovations.

What would happen if *all* agriculture went the way of Gabe Brown's farm? After we spent the day walking around his fields, we sat in a barn to talk and watch a slide show Eliav Bitan had prepared. Bitan had gathered up the meager scientific studies on some of these "future-friendly farming" practices. Even a modest two-or-three species cover crop causes a 90 percent reduction in sediment runoff and a 50 percent reduction in fertilizer runoff into the watershed, and sequesters a metric ton of carbon dioxide per acre. Bitan added together the measured impact of simple cover crops and other helpful practices—although none, in these studies, was undertaken to improve soil health—and surmised that these practices on US cropland could absorb 5 percent of greenhouse gas emissions in the United States. But he thinks that estimate falls far short of the real potential.

"What if you were actually trying to maximize the soil?" he asked. "Could the soil absorb a fifth of our greenhouse gas emissions? Some people think we can, but the science just doesn't exist yet."

How might healthy soils affect the annual flooding that goes on in North Dakota and other states? Floods cause around $8 billion in damage every year, and states spend billions to prevent them—North Dakota alone plans to invest nearly $2 billion in a 35-mile flood diversion channel around Fargo. But what if the rain penetrated the land, as it does on Gabe Brown's farm, instead of running off? It took him 15 years to create the kind of healthy soil that can absorb a very heavy rainfall, which is another commonplace of climate change. But now he has little fear of flooding. In 1993, an infiltration test of the soil in one of his fields showed that it could only absorb a half-inch of rain in an hour. A 2012 test showed that the same field can now absorb 8 inches in an hour.

Many soil scientists are stuck in their labs and research

plots, but Kris Nichols's job compels her to spend time with farmers and try to make the science serve their needs. Besides, she's deeply interested in what's going on out there. Her father was the commissioner of agriculture in Minnesota and her family has a long history with the challenges of farming. "My career was probably supposed to be looking at fungi," she said with a laugh when I visited her in Mandan. "A fungus caused the potato famine in Ireland and that brought my family to this country!"

Just as she has helped Gabe Brown change, he's helped her change, too. "Sometimes education can be limiting," she told me. "We think we can go just so far, but what Gabe and Marlyn and the others have shown me is that we can go a lot farther. They've shown me something unbelievable."

Nichols stands out among her peers, not all of whom are convinced that enough carbon can be captured and stored stably enough in the soil to make an impact on global warming. They look at the most intransigent part of soil carbon, called humin—the part that's been eaten and processed by microorganisms so many times that it's nearly inert—and note that it takes hundreds and even thousands of years for that sort of evolution. Some of these other soil scientists aren't impressed by the many other less-concentrated forms of carbon in the soil, because that carbon could more easily be lost to the atmosphere. But Nichols pointed out that farmers like Gabe Brown are adding vaster amounts of carbon to the soil than any of their agricultural predecessors did. And given their dense concentrations of cover crops, much of the carbon they add comes from root exudates, which become humin more quickly. In short, she said, these other scientists aren't familiar with the innovations of the Burleigh County farmers and other agricultural pioneers around the county.

"Some of the things we're doing with cover crops and livestock are adding so much more value than the old conservation methods," Nichols explained. "Will that restore the prairies' native carbon content? Personally, I think it's possible."

There's another reason I love the story of the Burleigh County farmers and ranchers. Most people think that good agriculture is a luxury reserved for boutique farmers who work a day job as computer programmers or whatever. Gabe Brown and his colleagues have shown that when you understand nature and work with her, farming becomes easier and cheaper, not harder and more costly.

"If we get out of the way, nature will do most of the work," Rick Haney told me. And that's the path of least resistance that Brown follows. He crisscrosses his fields with heavy machinery far less than he used to. He sprays hardly any chemicals. His two herds—cattle and microbes—clear up his fields. And he doesn't want genetically mollycoddled cows that don't know how to be cows. When he turns them out onto snow-covered fields, he expects them to nuzzle aside the snow and eat the cover crops buried there. He expects them to figure out how to meet their moisture needs with snow so he doesn't have to drag water out to them. If they don't, he sells them and invests in sturdier stock.

When the soil tour visited Marlyn Richter's farm, Richter regaled the crowd with stories of his parents' disapproval when he and his brother wanted to change the routine. He had been equally scornful himself only a few years earlier. The Richter brothers are third-generation farmers. His grandfather bought the farm during the Dust Bowl years. His parents raised 14 children there and still live there themselves.

"I used to see Gabe Brown with his soil charts and his talk about water infiltration, and I'd think that it was all BS," Richter told the group. A big guy, he was wearing a gym T-shirt, and his chest and arms suggested he went often. "I was tilling and I thought, 'No way that water's not going down!' But I was driven to change by struggling with these sandy soils." Then he waved his hand at one of the fields, which used to be so hard that it was tough to sink a plow into it. Now he's gone no-till, planted cover crops, and grazes his just-weaned calves on it. His soil has gone from 0.8 to 2.6 percent organic matter, and he's reduced his herbicides and cut his fertilizer use by 15 percent.

He and his brother are also working *less*. They now take Sundays off. That really shocks their father, who was so convinced of the need for grueling, grinding, never-ending work on the farm that he used to tell the boys to hoist an anvil and carry it around if they seemed to be slacking.

I repeated this story to Gabe Brown in his basement and he laughed. "I tease my buddy Marlyn all the time that he's not moving fast enough," he said. "They need to get rid of all the commercial fertility. They need to put those milk cows out to graze. They're doing things like chopping up feed and bringing it to the cattle. We used to do that years ago, too, but we want to let nature run its course. Let the cattle do the work."

He settled his cap back on his head. "Marlyn's got to get lazier."

CHAPTER 5

CASHING IN ON CARBON

I drove the nearly 2 hours from the Perth airport to Bob Wilson's house with a red ribbon tied around the pointer finger of my left hand. I waved it every time I had to make a turn or face oncoming traffic. "Keep to the left, keep to the left," I chanted in an effort to keep myself on the correct side of the road. "Be a leftist!" Honestly, I don't understand why the Australian government even allows Americans to drive there without taking a course. I was more frightened exiting the Perth airport parking lot than I'd ever been during my many weeks in Afghanistan working on a book (*Kabul Beauty School*), even more rattled than when I'd been driven through a part of the country rumored to be Taliban friendly. I figured my life was much more likely to end on the chrome snout of an oncoming truck in Australia than at the hands of some Talib chumming for small foreign fry. And there was the poor truck driver to worry about.

My fear didn't dissipate when I finally turned down the country road called Mimegarra that would take me to Wilson's farm. He had appended this note to his e-mailed directions just before I left the States: "If a kangaroo runs in front of you, keep going straight ahead. Do not attempt to stop or swerve!" I drove down the road so slowly that butterflies outpaced me, but fortunately, no kangaroos bounded from the bush at the sides of the road to make me test Wilson's order—given, he said, because I'd likely hit the kangaroo *and* flip the car if I tried to stop or swerve. I meandered along, taking sad note of the humps of fur and blackened hide from the many unfortunate collisions between kangaroos and cars. And I couldn't help noticing the white powder dusting the edges of the road, but didn't stop to find out what it was. Some kind of fertilizer? Ash?

I figured out what it was when I arrived at Wilson's farm and took the wrong road, up a hill lined with dying pine trees and past the driveway to his home. My car sank and my tires struggled, even though I was clearly on a road of sorts. I opened my door and looked down to see that the road was just a thin covering of grass and wildflowers over a slope of fine white sand. A green wig over a white skull.

"Yeahhh," Wilson told me later, after I backed down the hill, parked near his home, and was discovered by his Jack Russell terrier. "We're basically on top of an immense sand dune here. The ancient waterline was over there, at the Darling Escarpment." He pointed across a valley at the long green ridge I'd driven just to the west of on my way up from Perth.

Wilson and his wife, Anne, had offered to host me on a whirlwind tour of Western Australia's soil health activists.

Many of these activists had been influenced by Christine Jones, a former academic who studied sheep, became interested in the way the quality of their wool was affected by what they ate, and then became interested in how the health of the soil affected their forage. Back in 1981, when she was a research fellow at the department of agronomy and soil science at the University of New England in New South Wales, she began speaking to farmers about their soil and convincing many that the key to success laid in their underground trove of carbon. She was very specific about how to build that carbon, too, urging them to keep the land covered with growing green plants, which feed exudates to the soil microorganisms—what she calls the "liquid carbon pathway." After seeing a television program on Jones's work to rebuild landscape health, an Australian couple brimming with admiration—Allan and Kay Hill— offered her $125,000 to speed her work along. She used the money to set up a series of annual awards for agriculturalists who were doing the best farming for soil health as well as keeping their businesses profitable. Wilson won the first $25,000 A. and K. Hill Green Agriculture Innovation Award in 2009 and was going to spend the next few days taking me around to visit other outstanding farmers in Western Australia. Then, I would fly to the east coast of Australia and attend Jones's fourth award ceremony in Bega, a town in a bucolic valley near the sea in New South Wales.

Shortly after my arrival, Wilson whisked me outside to see his operation before the sun set. It was hard walking back to the car, as the sand sucked at my shoes where the ground was bare, and hard to imagine anyone growing anything on it. He told me that if he were farming conventionally—in this part of the country, that used to mean plowing and planting shallow-rooted annual grasses and clovers for the cattle to graze—he would have failed long ago. However, Wilson is an innovative farmer, like Gabe Brown, and began testing new ideas years ago. Back in 1985, he started planting a legume shrub called tagasaste to feed his cattle during the dry months and to keep the wind from whirling away his topsoil, such as it is. He now has about 2,500 acres of pasture criss-crossed with hedgerows of tagasaste.

Then in 2003, he took the far more radical step of plant-ing subtropical perennial grasses. Also like Gabe Brown, he had a partner in this innovation, an extension officer from the commonwealth's Department of Agriculture and Food named Tim Wiley. It was an anxiety-producing move, as they didn't know which perennials would perform well, if at all, and the seed was expensive. But the new grasses flour-ished. And because they were deep-rooted perennials, they could withstand the vagaries of Western Australia's increas-ingly erratic climate.

Back in North Dakota, Marlyn Richter had told the visiting group of farmers and Natural Resources Conserva-tion Service employees that farmers couldn't be dummies. "It used to be that all you needed for farming was a strong back," he said. "It was okay to have a weak mind. It's not like that anymore."

Just as Richter, Brown, and the agriculturalists around Bismarck educate themselves about soil health, so have these farmers in Western Australia. Wilson told me that because farming and ranching represent a smaller and smaller part of the GDP, the Australian government has cut back on funding for the agriculture department. So the farmers themselves have banded together into groups— Wilson is the head of one called Evergreen Farming, which created a cheeky bumper sticker (one is now stuck to my refrigerator) demanding SHOW US YOUR GRASS!—and seek funding for their own field trials. Four years after Wilson planted his perennial grasses, he and Wiley tested the soil carbon content from the surface to 3 meters down. There was more carbon under the perennial grasses and tagasaste hedges than under the annual grasses—not surprising, as the lack of tillage leaves intact the healthy soil communities and the perennial grasses and shrubs keep a growing root system in the ground year-round that feeds carbon sugars to the microorganisms. Wiley wrote up a paper suggesting that perennial grasses might play a role in sequestering soil carbon and mitigating climate change and—citing some of the data from Wilson's property—submitted it to a Senate inquiry into climate change and agriculture. They were asked to join Christine Jones in making a presentation, as were nearly 20 other groups, in 2008.

"Most of the other groups were talking about the problems that climate change was going to cause for agriculture and how they needed money for research," Wilson told me.

"They were all doom and gloom. We were the only group who said we thought we could solve the problem."

Wilson pointed out the remains of the pits that were back-hoed to test the soil's carbon—given his sandy soil, the carbon looked something like chocolate baked on to the top of vanilla sponge cake—and showed me some of his new trials comparing different organic soil treatments. Then we roared off in his Land Cruiser to round up some escaped cattle, the Jack Russell perched on his right knee. Wilson leases 5,000 acres on this sand dune and grazes 1,000 cattle, some of which he sends to Israel, where they are fattened in a feedlot and then sold in Europe. We zoomed past the tagasaste, which looked something like a formal hedge maze set in the middle of a pasture full of yellow daisies. As we neared the end of one field, Wilson stopped the truck and pointed at some creatures bounding along a fence. "Emus," he said with disdain. "Bloody stupid birds get caught in the fences and break up the wire." We passed through several paddocks—he moves his cattle through a series of pastures so that they will graze Savory-style throughout the year—and finally found the breakaway bovines. Wilson chased them back with the Land Cruiser, circling them and racing his engine when they dawdled. As Wilson succeeded in making one bull gallop away, a bull in another pasture yodeled with frustrated aggression.

Back at the dining room table, Wilson and I were talking when his wife, Anne, interjected from the kitchen. "Are you going to tell her your dirty little secret?" she asked.

He flushed, his sun-bleached eyebrows standing out even more than before in his tanned face. "Yeahhh," he said. "I no longer believe in climate change. That is, I believe the climate is changing—since the mid-1970s, we've really been dealing with another climate entirely here in Western Australia, where there's less rain overall throughout the year and more of that falls outside the growing season. But I don't believe that human activity has caused this."

A climate skeptic! I listened with only faint surprise as Wilson told me about Web sites he follows, like JoNova's from Australia and Anthony Watts's from the United States. He recalled things he's read that make him doubt the climate alarms: that the "hockey stick" graph of rising global temperatures fails to mention a warming period during the Middle Ages, which suggests that climate swings have occurred naturally throughout time. That ice core measurements show CO_2 levels rising after temperature hikes, not before, which suggests that atmospheric CO_2 doesn't promote warming.

I wasn't surprised by any of this, because people involved in agriculture are often skeptics of anthropogenically driven climate change. The American Farm Bureau is one of the most vociferous groups challenging climate change in the United States and claims that 70 percent of American farmers share its stand. When I talked with David Miller, a farmer and the director of research and commodity services for the Iowa Farm Bureau, about rewarding farmers for the public good of soil carbon, we delved for a bit into the agricultural community's feelings about global warming (I avoided talking about the science). "There was trouble from the beginning with how it was presented," he said. "There was the feeling that the scientists had been co-opted by the politicians and that affected their credibility. The closer someone is to the land, there is the sense that we've been through big cycles before. But that was absent from the global warming debate. What farmers understood was totally ignored or not addressed."

Bob Wilson was a believer when he testified to Parliament in 2008, but even though he changed his mind, it still makes sense to him to do everything he can to build his soil carbon and have the world take notice. First, he's seen how the buildup of soil carbon underneath his perennial grasses is helping the land hoard the rain falling on his sand dune, making it more productive and preventing erosion. That has downstream benefits for the environment around his land, too. And second, Australia may become the first country to pay farmers for building up soil carbon as part of its effort to combat global warming. Wilson may be a skeptic, but he's not about to turn down cash for this other "crop"—one that's beneficial for his farm and for the public—that he's been tending so very carefully over the years.

Beginning in 2012, under Australia's Labor government, the Clean Energy Future plan slapped a carbon tax of $23 per ton on the country's greatest emitters—energy providers, transportation companies, and other industries that churn out lots of greenhouse gases—but does not go after agriculture, which is generally assumed to create 13 to 14 percent of the world's emissions.* In fact, the country's Carbon Farming Initiative (CFI) offers agriculturalists a chance to claim some of these carbon-tax dollars with activities that store carbon or reduce greenhouse gas emissions on farmland. When I was visiting Australia in September of 2012, the only accepted method for biologically storing carbon was to plant trees and pledge not to cut them down for 100 years. This rankled many farmers, who thought that it made no sense to turn working farmlands into forests when one of the world's other looming crises is feeding a population projected to grow to 9.6 billion by 2050. Instead, they hoped that the CFI would expand the list of remunerated carbon-storage activities to include building soil carbon with better land management, which farmers around the world had already been calling carbon farming—at least since 2006, when Vermont's Abe Collins cofounded a company called Carbon Farmers of America.

Carbon Farmers of America offered to sell credits called carbon sinks—a ton of atmospheric carbon dioxide converted to about 545 pounds of carbon in the soil, newly relocated from the air to the soil by farmers—to anyone wanting to invest in soil renewal and perhaps offset their own carbon footprint. Collins's group also envisioned a marketing plan to certify their products, ensuring customers that in addition to

---

* A newly elected and more conservative government threatens to repeal the tax, perhaps by July of 2014.

being nutritious, their food was helping with the climate crisis. It ultimately wasn't a workable plan, but it was an inspirational one.

But when there's a possibility of real dollars changing hands—"real" as in the $1.7 billion from its carbon tax revenue that the Australian government planned to spend on its Carbon Farming Initiative—figuring out payments for soil carbon sequestration has been slow and controversial. Some of this has to do with the very opacity of the soil. No one doubts that a growing tree removes carbon from the air and locks it up in a solid form: There it is, right in front of your eyes, whereas carbon in the soil remains hidden. Since understanding of the soil has generally lagged behind that of other ecosystems, convincing policy makers and others that the soil can absorb and hold carbon is a tough task.

Since the benefits of building soil carbon are potentially huge—resulting in not only more productive farms, cleaner waterways, and overall healthier landscapes, but also global warming mitigation—some general rules have been developed to align this biological accounting with an economic one. Even when all parties agree that each ton of carbon stored in the soil is equivalent to about 3 tons of carbon dioxide in the atmosphere, any plan to reward activities that store carbon in the soil has to consider several factors to make that storage meaningful. First, the amount of carbon stored has to be measurable in order to fit into an accounting system. Second, the activity that stores the carbon has to be additional, meaning that the agriculturalist undertaking no-till or cover-cropping is new to the practice—a condition that sadly leaves pioneers and early adaptors like Gabe Brown and Bob Wilson in the lurch. Third, the activity that builds soil carbon can't cause leakage, meaning that it can't create a corresponding decrease in soil carbon elsewhere. Say an agriculturalist undertook some sort of carbon-building practice that made his or her land less productive, prompting another farmer or rancher to make up the difference in supply using conventional methods that destroy soil carbon and release it into the atmosphere—that would be leakage. In other words, there would be no overall increase in soil carbon.

Fourth, the carbon storage has to have a meaningful lifetime, an idea that has caused some heated debate around the idea of permanence. For instance, a farmer might build up the soil carbon on her land for 5 years, but if she sells the farm to someone who turns it into a housing development, that accumulated carbon will puff away in the aftermath of the bulldozing. Even the most stable forms of soil carbon will break down in a few years after they and their microbial and fungal communities have been sliced and diced, just as the carbon stored in a tree is released into the atmosphere by fire. Reasonably, someone paying for soil carbon sequestration wouldn't want that to happen, but the fact is that agricultural lands change hands.

"Trying to fit in the idea of permanence makes the program unworkable," David Miller told me. "A significant portion of agricultural land operates on one-year renewable leases. Our experience is that in a five-year period, about 17 to 20 percent of the land will change operators. That's a lot of movement."

Difficult metrics to apply to a complex biological system where few things are permanent! Even coal is not a permanent form of carbon storage, of course. Perhaps only diamonds are. But smart people around the world are working hard to develop elaborate, scientifically sound protocols—for either the voluntary or compliance markets—in which soil carbon storage and other forms of greenhouse gas mitigation can be verified and rewarded.

The voluntary markets allow corporations and other entities that want to please customers and shareholders with some green action to buy credits from practitioners that either reduce or prevent greenhouse gas (GHG) emissions or sequester soil carbon. A handful of nonprofits develop the protocols on which these credits are based, putting together the best research from peer-reviewed studies on the effects of a certain practice, writing up comprehensive descriptions of what the land managers have to do, then verifying every year to see if they're are following the program. As of 2012, the voluntary market in the United States had issued more than $100 million in carbon credits.

The compliance markets are driven by government regulation of greenhouse gases, which require large GHG-discharging entities to reduce their emissions or buy credits that offset their emissions. Significant compliance markets include California's cap-and-trade program (which was just getting started in 2013) and the European Union's Emissions Trading System. The California program has been taking protocols developed for the voluntary market and fine-tuning them to fit the state's regulation.

Many progressive ranchers and farmers are stepping forward to help develop protocols. "They're coming to the table," explained Belinda Morris, formerly of the Environmental Defense Fund and the Nature Conservancy and now California director for American Carbon Registry, a nonprofit that develops protocols for the voluntary markets, registers carbon projects, oversees verification by third parties that agriculturalists are actually preventing greenhouse gas emissions or building soil carbon, and issues credits for sale. "Many of them think they're eventually going to be regulated anyway, and they want to have a say," she continued. "For instance, the California rice industry has been working on a methane reduction protocol with us because they think regulation will come, and they want to see it done right. They also want to be able to take advantage of new markets for additional revenue as opportunities arise."

Even outside the opportunities specific to global warming, healthy soil is becoming an item of interest. "There are a whole lot of groups out there that are doing sophisticated analysis on a large scale of the value of having healthy soils and healthy landscapes," agricultural and natural resources economist Sara Scherr told me over the phone. "I think that's going to seep into policy and that people are going to become more aware of the cost of ignoring that kind of land management for a variety of ecosystem services."

"Ecosystem services" may sound like the name of a waste-management company, but that's the term used to trumpet the idea that healthy, carbon-rich soil is a win-win for just about everyone (with the likely exception of

Monsanto and other purveyors of GMO and chemical agriculture, should the regenerative model of agriculture prevail and turn them into this century's buggy whip manufacturers). I first heard Abe Collins speak at the Quivira Coalition's conference in 2011, and he may be the nation's most stirring voice arguing that soil carbon is the most basic and important life-supporting infrastructure on earth. He rattled off an impressive list of ills that healthy soil can cure. It not only keeps water in the soil, but also allows the microbes to gnaw away pollutants as this corralled water slowly drips into streams and aquifers. That water storage mitigates flooding and wildfire, and the soil aggregates themselves prevent both water and wind erosion. That means fewer particulates dirtying up the air, so healthy soil brings cleaner air, too. Plus, healthy soil is highly productive soil. "This is an opportunity for farmers," Collins declared to the crowd. "We know how to build topsoil. We can fix land, not just preserve the degraded landscape that's already in place."

I talked to Collins frequently by e-mail and phone over the next few years and finally visited him in Vermont the week following my trip to North Dakota. A few minutes after I checked into my hotel on my first evening, he knocked on my door and asked which I would prefer: resting up and meeting in the morning, or drinking gin and tonics on the buddy seat of his tractor while he sprayed his fields with a mixture of liquefied fish and minerals.

"And raw milk?" I asked. I knew he'd also taken to

spraying the leftover skim milk from his dairy creamery operation—which doesn't fetch much of a price—onto the fields. An extension agent in Nebraska named Terry Gompert and some farmers there found out that the soil microbes love it. New insights like that spread quickly through the world of regenerative agriculture.

Collins shook his head. "No milk." Then, "Sorry I smell like fish. But it's good to spray this stuff right after it rains and in the dark."

I followed him out to the farm he was building as the green hills turned black and clouds curdled in the sky. Collins had a bar of sorts set up by a barn. He cut up some limes on a rough board poised on sawhorses and poured Hendrick's gin and some tonic into two Mason jars, dubbed it a fine example of "Ginny Hendricks," then left me to nurse mine while he struggled with the rigging behind his red tractor. There were two small lights on the front of the tractor and its fork was lifted, making it look like a lobster in attack mode. "I really appreciate the low-input way of farming, but sometimes you just have to fire up the machine," he shouted over the noise of the engine.

Like Gabe Brown, Collins is another farmer who doesn't come from a farming family. He lived in the Hard Rock Chapter of the Navajo reservation in Arizona back in the 1990s, working on a locally led project to harvest water and reclaim desertified lands. Young men were using techniques from permaculture, installing swales to harvest water, plugging up gullies with gabions, and planting trees.

Some of the older Navajos told him that their land was healthier when they were more nomadic, grazing their herds and moving them on and not returning to the same spot until the grass grew back. Later, he would hear that idea echoed in Savory's belief that landscapes in brittle environments depend for their health on intermittent but intense grazing by large herds behaving as they did in the presence of predators.

Collins returned to Vermont and began farming, putting what he'd learned out west to work. He was always as interested in soil health and overall landscape function as in the daily challenge of keeping feed in front of his cattle. He read deeply into the agricultural literature—P.A. Yeomans, André Voisin, Sir Albert Howard, Robert Rodale, Allan Savory, Louis Bromfield, Friend Sykes, Newman Turner, and more—and talked with farmers from around the world at conferences. He tried their grazing tweaks and some of his own to improve the farm. He realized he was making a real breakthrough when he dug some postholes and found 16 inches of good black soil covering the blue Vermont clay, not the 8 inches he'd had the year before.

While Collins wrestled with the rigging behind the red tractor, he talked about his newest way of making a living building soil carbon. He had created a business called Collins Grazing that builds healthy farms from the soil up—since the primary infrastructure on a farm or ranch is the soil, every component therein has to contribute to its health—for clients with the capital to invest in tired farms.

For instance, on the farm he was building when I visited, they had invested in the lightweight electric fencing that allows holistic grazers to create many small pastures and move their herd of cattle from one to the next just by opening up a gap. One person and a few miles of electric fence can handle a herd with as much finesse as Savory's gang of herders. Through Collins, the owners also invested in cattle, enough to affect the perfectly timed impact of many animals on the land. Typically, Collins's customers can afford to wait until the land makes its transition from a depleted conventional farm to a carbon-rich, productive one.

"I can build them a bunch of soil," Collins said. "That's a new business plan for me: creating turnkey farms from the soil up for my clients, working landscapes with the most advanced ideas from regenerative agriculture. This is expensive land, but it's going to take a lot of work to get it to the point where they can make money on it."

Putting together boutique farms might be a good way for Collins to make a living, but his vision for soil health is far from limited to these particular acres. As we talked throughout the next day, he explained that cities spend billions on downstream problems from depleted landscapes. They have to put in infrastructure to avoid flooding, dig the agricultural erosion out of their ditches and waterways, and put their drinking water through a maze of treatments to make it potable. "Paying land managers to rebuild soils is just smart urban planning," Collins said.

Months later, he sent me the link to a Forbes.com blog

post that made exactly the same argument. In it, reporter Steve Zwick pointed out that by 2012, 200 cities in 29 countries had decided to forego building new water treatment plants and reservoirs and, instead, invested in watershed restoration that reduced pollution downstream—double the number of such efforts in 2008. Zwick cited a report from *Ecosystem Marketplace* that New York City alone dodged spending $6 billion on a new filtration plant in the 1990s. Instead, the city employed a number of inexpensive strategies to improve the water supply, including paying Catskill farmers upstream to change their land management practices in ways that reduced runoff on their properties and, ultimately, pollution in the lakes and streams that provided the city's drinking water. Not only did this investment in upstream farmers save vast sums of taxpayer dollars, but this "natural" water purification system continued to serve the citizens even when Hurricane Sandy roared ashore and zapped the electric power. According to Zwick, there are 67 similar programs in the United States. Worldwide, more than $8 billion is spent to improve watersheds' natural function in order to protect water quality downstream, with China spending 90 percent of those dollars.

So every time Collins observes that his and other farmers' management of the land has resulted in healthier soil—with the attendant benefits of greater productivity, better water filtration, less erosion and runoff, and more biodiversity—he can't help but imagine how the watershed would be impacted if a thousand other farmers were doing what he does. And he can't help thinking about the urgent need to rebuild watersheds. "It's a biological process of working with plants to turn air and water into the organic matter that turns subsoil into living organic topsoil," he told me. "How quickly can that be done? I think the jury is out on that, so we need to monitor and be open to the out-liers and the anomalies"—like spraying raw milk. "But definitely, slow soil formation isn't going to get us out of the corner we've painted ourselves into. We need to regenerate it on a pretty tight time frame."

Anyone expecting Collins to be some sort of nostalgic Luddite quickly has that expectation adjusted: He loves technology. For several years, he's also been contributing to the development of a Web-based decisions-support tool for farmers that crowdsources innovation and spreads it around the globe. "The burning of the libraries at Alexandria is considered a great historical loss of information," he said. "But to me, there's a comparable one happening every day, the loss of information generated on farms and ranches. People uncover so much, there's a spark and a flare and maybe someone writes it down. But information technology gives us the chance to link around the world."

Collins is also eager to see high-tech monitoring of working landscapes, an interest piqued when he and his part-ners in Carbon Farmers of America needed to demonstrate the accrual of soil carbon on their lands. Such monitoring will not only give farmers and ranchers another way to gauge the effectiveness of their land management, but also demonstrate that soil carbon can be increased to others who might need numbers to be convinced.

Soil carbon is typically tested by removing cores from within a plot, then drying and analyzing them. There is one huge problem with this technique, and that problem has triggered criticisms of the idea of paying agriculturalists to build soil carbon. The problem is that landscapes comprise huge variability—different concentrations of sand, clay, and silt, different topographies, different patterns of usage by humans and animals, different weathering—and the soil carbon content in one core may be quite different from a core taken 5 feet away. Even within one plot, the reading may vary up to 2 percentage points—a huge amount when you're talking about soil carbon. Fifteen years ago, a soil scientist named Dan Rooney, who had specialized in graduate school on miniaturizing geotechnical sensors, figured out how to combine proprietary sensors with software to accurately map the soil properties of entire landscapes. "I can not only give someone information on the way their soil is right now, but I can tell them what they can expect it to become given a certain kind of management," Rooney told me. "That's important, because it's all about the site's capacity to give us more."

Rooney's technology is mostly used with high-value crops like grapes as well as in what's called precision agriculture: large-scale operations that try to minimize the amount of irrigation and fertilizer a crop might need based on a detailed understanding of what's happening in the soil. He's used his technology in China, North Africa, Europe, and the United States, helping industrial agriculture be more efficient and less wasteful. But he also worked with Collins to get a baseline soil reading for what may be the world's most monitored organic farm.

The Innisfree Village farm is in the foothills of Virginia's Blue Ridge Mountains and, since 1971, has been a home for adults with intellectual disabilities. Part of the village's founding mission—stemming from a belief that living on a healthy productive landscape is good for people—is to conduct environmental research. In 2011, Innisfree hired Collins and Rooney to do a carbon baseline on 62 of their 280 acres of pasture. The community was curious about how their management of the farm's cattle, sheep, and chickens might be changing the soil. That bit of information made them hungry for even more data about the ways in which their land was changing. They discovered a worldwide "observatory" movement focused on sensing infrastructures that detect environmental change and then feed that information into software that aids analysis and decision support.

Innisfree farm manager Peter Traverse took the idea of turning the farm into an observatory to private funders who are interested in healthy landscapes, and the Innisfree Observatory was born. Along with the more common tools of farming, the staff there deploys probes that detect volumetric moisture in the top 4 inches of the soil and penetrometers that measure soil compaction, both of which are tied into a GPS system. They use iPads to take photographs and evaluate the amount of bare soil whenever they move the cattle. They have a stream gauge that measures water quality as well as the volume of water crossing the property via six streams, and a 3-D anemometer that monitors the wind. They're also working with a scientist from Montana State University on developing a sensor that will monitor the ongoing carbon level in the soil.

"We're engaging the land in an active conversation," said Traverse when I called him. "There are now ways to ask the land how it's feeling. Of saying, 'I just did this, and how did it work?'"

The goal of the Innisfree Observatory is to connect land managers like farmers with scientists, so that the scientists have a connection to an ongoing examination of how different land management strategies play out on large tracts of land in real life—as opposed to on small research plots—and so that the farmers and ranchers learn to collect and publish data that's of value to the scientists. All of this has been set up with the input of a science board to ensure that when Innisfree Farm and the area farmers that join the observatory collect and publish data, it will be considered to be of research quality by any scientist anywhere—not to mention useful to other farmers. The portal to the site was built by people from the University of California–Berkeley's environmental engineering department.

"We want legitimacy," Traverse explained. "Most of us who are doing regenerative agriculture really feel strongly that we are making a difference, but there's a substantial lack of documentation to convince the rest of the world.

Our project is trying to provide an infrastructure that might allow that to happen."

Traverse expects that other farmers can one day avail themselves of some of the fancy sensors he's got in place now. He believes they're consistent with the classic observational skills of farming, "I'm from a multigenerational farm family, and I really enjoy watching my animals and the change on the landscape because I learn all the time from that," he said. "I also like watching what my soil biology does and what the water is doing. I couldn't see that before, and neither could my father and grandfather."

The Innisfree Observatory is a great example of what Abe Collins wants to promote—high technology close to the soil—but he's also involved in a very low-tech effort to measure the impact of land management on soil carbon. In 2011, he and several soil health advocates launched the Soil Carbon Challenge, described on its Web site as an international competition "to see how fast land managers can turn atmospheric carbon into soil organic matter. . . . If you want to find out how fast a human can run 100 meters, do you build a computer model, do a literature search, or convene a panel of experts on human physiology to make a prediction? No, you run a race. Or a series of them. There's been tons of talk about soil carbon, but it's time for motion: to show with good data what's possible, and recognize those land managers who know how to increase soil carbon."

For years, farmers and ranchers had been talking about the need for hard data to quantify the achievements of regenerative agriculture. The Soil Carbon Challenge was finally kicked off when a few of them attended a conference and heard a government scientist say that he didn't think land management made much of a difference to soil carbon storage and that it was too hard to measure the carbon accrual anyway. Organizers of the challenge suggest that it's something like an "X Prize" for farmers and ranchers, although the award dollars come not from a billionaire but from fees paid by the participants. The hope is that this grand quantification will prod governments and institutions to develop incentives to encourage carbon-rich farming and grazing. Perhaps this careful monitoring of change could even lead to soil carbon being at the heart of the US farm bill and our public land policy. How much could the planet change if that kind of official embrace were global?

Such an audacious project, such a big idea! And sort of amusing, to me, that the architect and conductor of the Soil Carbon Challenge is one guy living in an old yellow school bus.

I went to visit Peter Donovan shortly after I moved to Portland in the fall of 2012, right around the due date for my second grandchild (in fact, I had to make sure a surrogate grandmother was around to watch baby number one if my daughter and her husband had to bolt to the hospital while I was out of town). Donovan had parked his bus at a farm outside Philomath, Oregon, not far from where pigs lounged in the shade around open-air hutches in a large field and where a vigilant field marshal of a dog was driving sheep through a chute to be weighed. Pots of herbs soaked up the sunlight on the bus's hood. Donovan had turned the inside into a comfortable living space fancied up by an upright piano bolted to the floor. At one point during our interview, he played me a thunderous classical piece and I couldn't help imagining him there at night, tinkling the ivories by star shine, with only the pigs and sheep and wild creatures to listen.

This makes Donovan sound like a hippie. But he's somber, even severe looking, with short gray hair and triangular black eyebrows and no-nonsense crinkles around his eyes. He has the gravity of a surgeon, but one who cuts his own hair.

Donovan has committed to life on this bus for at least 10 years, which is also the level of commitment he expects from participants. At the beginning of the 10-year cycle, he takes a baseline soil carbon measurement from a spot on the property where the farmer or manager has not yet undertaken new management—say, no-till or holistic grazing or cover crops or any of the approximately 40 broad strategies that Abe Collins says regenerative farming employs and adapts. The spot is always one that the farmer or rancher thinks has potential and has plans to begin new management there after the measurement. Donovan will return every 3 years to take a new measurement. The project is not intended to gauge the carbon increase over the entire property, as one would need to do to measure and sell credits for tons of carbon sequestered. Instead, he wants a snapshot of the kind of change that creative and dedicated land managers can make on soil carbon levels.

That 10 years is one of the things that sets Donovan's effort apart from past soil carbon measurement projects. Most academic research projects, he said, have to be completed and reported within 3 years, which isn't enough time to track what's happening. And many estimates of how much carbon can be stored in the soil are done by comparing different landscapes to one another: They measure the carbon in undisturbed forest soils, in soils that have been farmed for 20 years, in soils that have been farmed for 10 years, and so on, then calculate how changes in any of these directions appear to have affected soil carbon levels. But by monitoring the change at multiple sites over 10 years, Donovan can really pinpoint the impact of human management on soil carbon.

Most of the policy-level thinking about how to reward land managers for building soil carbon is based on modeling. The longest-lived example was the Chicago Climate Exchange, which ran from 2003 to 2010, launched amid optimism that the United States would adopt a national cap-and-trade program. The exchange based payment on models, basically saying that certain practices like no-till or manure spreading equaled a set amount of carbon stored in the ground. California's cap-and-trade program has a similar approach, although its protocols are far more rigorous, conservative, and science-based.

But Donovan believes that modeling is basically wrong-headed. Actual monitoring—whether by the slow, low-tech painstaking method of taking many core samples and sending them to the lab or by a high-tech method like Dan Rooney's or the kind of ongoing measurement happening at Innisfree—is the far better approach, he said. "Modeling offers a recipe and it limits the creativity of the land managers in the field," he said. "You're not going to get a guy like Gabe Brown"—who is participating in the challenge—"to do the stuff he does on the basis of paying for practices."

Donovan took me out to a site where he had done a baseline measurement—a field where turf grass had once been grown and then sliced out for replanting into people's lawns. Since his measurement, the farmer had put in perennial grasses and was now grazing his sheep with holistic management. Ordinarily, setting up a grid of sample points and pulling soil cores takes Donovan 2 hours, but he quickly explained how he went about it, then used his handheld tool to pull a 4-inch deep core of soil from the earth. He pointed to the yellowish bottom of the plug. "See how dry that is?" he said. "They're not getting enough water infiltration."

Then we hung around the bus for a while. This farm was not owned by the farmer who worked there; rather, it was owned by a group called Farmland LP, which buys old, tired conventional farms and turns them into healthy organic ones. A tour of investors was on its way and I wanted to tag along, but they didn't show up and didn't show up and finally, Donovan and I went to the farmers' market in nearby Corvallis (and there, I found the apple of my dreams—Ashmead's Kernel—and now have a tree growing in my yard).

But I was interested in the concept, since it was another example of someone building soil health and making money by doing so. When I caught up with Farmland LP's cofounder and chief scientist Jason Bradford on the phone, he explained that Farmland LP is a private equity fund that plans to take advantage of the gap between the supply and demand of organic food. Sales of organic goods have grown by 20 percent per year since 1990 and exceeded $28 billion in 2010. Even though organic goods are more expensive, the biggest limit to market growth is supply. Less than 1 percent of US farmland is certified organic. That acreage was increasing at 8.5 percent per year until 2008, but couldn't catch up with the demand; now the number of organic acres is in decline. Much of the demand for organic goods is satisfied by imports.

"Management and finance has been the bottleneck," Bradford told me. "There's a transition period of about 3 years between conventional and organic, where costs go up and income goes down. After that point, organic farms tend to do better financially. If you want to buy a car, you can get financing for five years. But right now, our system for financing in agriculture is short-term and pretty conservative."

Bradford saw that those 3 years of transition were a good target for finance, especially given that banks are leery of lending to organic farmers. He and his business partner, Craig Wichner, figured they could round up investors to supply the capital for land purchase and improvements. They would then be able to lease improved and certified-organic farmland to smart young farmers who had some experience with good farming but not enough money to buy their own places. Bradford and Wichner envisioned farms that would be not only certified organic, but also focused on soil health in a way that most of the large-scale commercial organic operations do not. They started with 150 acres in 2010 and now have 6,300 acres and 83 investors. "We tell our investors that they won't get their dividend payments for three years," Bradford said. "It's called patience capital."

Like Abe Collins, Farmland LP is tapping into a relatively small pool of wealthy people who think agriculture that builds healthy soil is not only a social good but also one that can be profitable—the old doing-well-by-doing-good approach. But there are probably not enough of that sort of wealthy person to save the 99 percent of US agriculture where land managers routinely destroy soil with poor practices. Who else can offer them the incentive to change?

We can, with the tax dollars that flow to the USDA. Founded in 1862 and signed into law by Abraham Lincoln—who called it "the people's department"—the USDA is

more often criticized than praised by good food and agriculture advocates. They cite the revolving door for the leadership of agribusinesses and the department, the USDA's lax
oversight of genetically modified organisms (GMOs), its
paltry support of organic and local agriculture, and the ways
it squashes small producers with rules intended to regulate
huge ones. As farmer and author Joel Salatin says in one of
his books, *Folks, This Ain't Normal,* "President Obama's secretary of agriculture, Tom Vilsack, was voted Governor of
the Year by the transgenic-loving industrial agriculture sector when he was governor of Iowa. Is it any wonder he would
approve transgenic alfalfa, more corn, and sugar beets? As of
April 1, 2011, eighty-one transgenic crops have been approved
by the USDA, and not one single request has been denied.
Who in their right mind thinks these people should be in
charge of food safety?"

But the USDA is a huge organization with more than
100,000 employees, and many are earnest strivers for agricultural policy that treats healthy soil as a public good. I
met one of these employees who works out of Portland last
year, and he explained how the USDA is supplying public
funds to help conventional farmers weather the transition
to a more sustainable agriculture.

Adam Chambers is a scientist with the USDA's Natural Resources Conservation Service (NRCS) and a farmer
himself, with a small property back in Kentucky—in fact,
he learned about the NRCS when he wanted advice for
transitioning his land from the single species of grass

planted by a previous owner to mixed perennial grasses that would enhance carbon sequestration on his land and offset his family's carbon footprint, as well as provide a better habitat for bees and other pollinators. "I did it mostly thinking about the carbon, because I work on climate change science and policy, and I wanted to walk the walk." he told me when I met him in the NRCS's Portland office, sitting near a wall of windows so dazzling with a view of the mountains that it was hard for me to focus. "But the song sparrows don't know that. The bobwhite quail don't know that. The turkeys that nest there don't know that. The water quality in the Ohio River—that knows nothing about carbon sequestration. These are public goods we're getting from this activity."

Chambers is on the service's Air Quality and Atmospheric Change Team, and he works on its Conservation Innovation Grant (CIG) program. Nine CIGs have protocols that either reduce greenhouse gases in agriculture or build soil carbon, provide farmers and ranchers with funds to make the transition, and then help them line up customers who will buy carbon credits for the avoided or stored carbon.

For one of these projects, the NCRS has partnered with Ducks Unlimited, a venerable organization of hunting enthusiasts that got its start during the Dust Bowl era—a catastrophic time for bird populations—and whose slogan is "Filling the skies with waterfowl today, tomorrow, and forever." Both the NRCS and Ducks Unlimited are dismayed by the amount of undisturbed land that's being plowed up and planted with commodity crops like corn, the price of which the ethanol market is driving the price sky-high. This eliminates nesting and breeding grounds for wild water fowl and destroys soil carbon. Fortunately, they've found a solution that pleases both conservationists and farmers and rewards farmers at the same time.

This CIG focuses on private lands in North Dakota that are part of the Prairie Pothole region, where the land is packed with thousands of shallow wetlands. These lands have been sheltered by the Conservation Reserve Program (acronym overload: CRP), a federal program begun in 1985 to pay farmers and ranchers to voluntarily rest some of their property from any form of production like row crops, grazing, or haying. Currently, some 27 million acres nationally and 3 million in North Dakota are locked up in 10- to 15-year CRP leases. The program isn't purely a conservation initiative, although it does protect land from the plow. The USDA likes to have a supply of sheltered acres in case some national emergency—such as the drought of 2012—threatens the nation's supply of feed for cattle. In those situations, the CRP lands can be grazed or mowed for hay.

But as the CRP leases expire, many landowners are gravitating toward the allure of high profits from commodity corn. So Ducks Unlimited and the NRCS have developed a third "working lands" alternative. In this scenario, landowners can choose to graze or hay the lands, but can't till them for crops—thus preserving the carbon that's been built up in the soil. The landowners need to develop a sustainable grazing plan, and the NRCS will help fund the investments those plans require, such as fencing and water for the animals. And brilliantly, Ducks Unlimited has lined up a major American manufacturer that will buy carbon credits from the landowners.

For decades, environmentalists have been nearly united in their belief that cattle ruin the land, but the science shows that waterfowl and cattle flourish in the same space. "The duck–livestock connection is pretty basic," explained Randal Dell, a Ducks Unlimited biologist (now working for the Nature Conservancy) whom I met in North Dakota. "Both need water or wetlands and grass. These Prairie Pothole wetlands are vital for waterfowl, providing forage and habitat at key life history stages. And in the eastern Dakotas, where Ducks Unlimited focuses much of its work, we're on the humid side of the Great Plains precipitation gradient. That additional precipitation and the plant community it supports provide plenty of forage that is pretty resilient to grazing."

In fact, Dell pointed out in a follow-up e-mail that these mixed-grass prairies evolved under the influence of bison and fire and require some disturbance to remain vibrant. He reminded me about how Gabe Brown's holistically grazed fields looked so much healthier than nearby CRP lands that had been ungrazed and undisturbed for years.

The only catch to the new third alternative is that participants must agree not to plow and crop this land—forever. They have to encumber their descendents and future owners with preserving this land for grazing and haying, never crops—meaning that they will never be able to reap the big ethanol bucks. However, what *sounds* like a stumbling block isn't, because landowners are scrambling to get into the program, Chambers said. NRCS set aside $1.5 million for the program in 2012, but there was $11 million in demand. "The public benefits of this pretty modest investment are air quality, carbon sequestration, water quality, wildlife habitat, and soil health," Adam Chambers told me. "And more people want to do it than we have public funds for."

While there's no controversy that tilling the soil releases soil carbon, there is still uncertainty in official quarters about whether land management can build soil carbon and by how much. Another CIG greenhouse gas project set in the Palouse—a beautiful area of rolling hills in Oregon, Idaho, and Washington—addresses this issue. Formed of silt that blew in from the west and cohered during the ice ages, the Palouse looks like miles of the kind of soft peaks you get when you beat egg whites, only very green. Wheat farmers have prospered there for generations, tilling their fields over and over—sometimes making eight passes with their equipment in a season. The soil carbon is not only getting pulverized and released into the atmosphere, but the soil itself is also eroding and causing problems in the watershed. Every time it rains, a torrent of chocolate milk–colored runoff pours from tilled fields into the streams. So NRCS has come up with a CIG that encourages farmers to transition to no-till agriculture, paying them $22 per acre for 3 years to help them with any extra costs or production losses. "It's a risk hedge," Chambers told me. "We actually think their production will gain."

The Palouse farmers will gain in another way, too. Scientists take a baseline soil carbon measurement down to a meter on participants' land and will measure again in 10 years. The farmers receive payment immediately for the soil carbon they're protecting from the ravages of tillage from EKO Asset Management Partners, a company that "invests in projects and companies that create environmental value." At the end of the 10 years, the Palouse farmers may also be paid for the amount of carbon they built in their soil over the decade. Several investment companies have expressed interest in buying carbon credits from the Palouse farmers in the future. In fact, the farmers will "grow" carbon just as they grow wheat: as a product they can take to market.

All of this makes my head spin a bit, but Adam Chambers tells me to think of the carbon credits as just another commodity—like an apple. EKO and the other investment companies will buy these apples in bulk from the Palouse farmers and then, at some point, will probably turn around and sell them in the voluntary marketplace. As of this writing, only six protocols have been approved for the newest compliance market, the California cap-and-trade program, and none have anything to do with soil carbon. Following along with Chambers's metaphor, that program is only interested in a very specific color, shape, and size of apple right now and is understandably leery of rotten ones. But there are huge opportunities to sell these credits in the voluntary market, where companies—often at the insistence of shareholders—are looking to boost their sustainability profiles.

"There is a growing interest in agricultural carbon credits," Chambers told me. "Purchasers seem particularly fond of the co-benefits associated with these credits."

Again, this CIG in the Palouse requires farmers to make a permanent transition to no-till. And again, the response from farmers has far outstripped the public funds allocated to the project.

Imagine if soil health were a bigger priority for our government—a huge priority, one in which ample funding was available for all agriculturalists to transition to sustainable agriculture (which Peter Donovan describes as "ruining the land more slowly") or, better yet, to Gabe Brown–like regenerative agriculture! Adam Chambers calculates that of the 914 million acres of farmland in the United States, only 4.3 percent is enrolled in some kind of government land conservation program. That leaves 875 million acres that could be enrolled in programs like the ones in North Dakota or the Palouse. Certainly, the public outlay would be immense, but the money saved on public health and on cleaning up our water, air, and the aftermath of climate-related disasters would dwarf the expenditure.

Australia's plan to pay farmers to store carbon or reduce greenhouse gas emissions not only eclipses national efforts in the United States monetarily, but in 2013 Australia also took the powerfully symbolic step of appointing one of its most eminent citizens as its first official advocate for soil health. Michael Jeffery is a career soldier who has since 2003 been the country's 24th governor-general, meaning he was appointed by the queen to serve as her representative in Australia. Ironically, his environmental philosophy derives, he told me, from Franklin Roosevelt, who signed soil-conservation legislation into law in 1936 with the words. "The history of every Nation is eventually written in the way in which it cares for its soil."

Jeffery retired from the army in 1993 and then served as governor of Western Australia unil 2000. Upon leaving that post, he founded a nonprofit called Soils for Life, which encourages regenerative agriculture and collects case studies from farmers and ranchers who have made the transition. In his new role as soil advocate, he's insisting that Australia look upon its soil and water as its two primary strategic assets, which should be managed together. He wants farmers and ranchers to be rewarded for improving the health of the landscape. And he's challenging the nation's scientists to study the attributes of healthy soil. "I want the scientists to give me the answer to a few simple fundamental questions such that we can't argue about them anymore," Jeffery told me. "Carbon sequestration is one of the questions."

I visited several of the farmers whose case studies are on the Soils for Life Web site, plus several that weren't. Most seemed doubtful that they'd ever be paid for the carbon in the soil, but the other benefits made the transition worth it. Their soil now holds water and minerals, and they're making money from that. Their streams run clearer, their animals are healthier, and they're tickled that scientists are now taking an interest in what they're doing. Farming itself is now more interesting, more creative—an art and a science, not a recipe.

"We're doing what we really love," Rob Rex told me. He has a sheep and grain farm near Wagin in Western Australia. "When we go into the local club and talk about farming, everyone else is all gloom and doom. But the people we meet on *this* trek are great. You can just see the energy in them."

CHAPTER 6

WHY DON'T WE KNOW THIS STUFF?

When I took a seat on my first day at the 2010 Quivira Coalition conference, my first thought was that I had never been in a room with so many cowboy hats. Soon I was dazzled by more than headgear.

In the morning, Doug Weatherbee from Dryland Solutions—he's an Elaine Ingham protégé, trained at her Soil FoodWeb center in Oregon—paced the front of the room to discuss his work helping poor farmers in Mexico improve their yields without expensive inputs like fertilizer, pesticides, and other chemical tools. He built a contour ditch system at a test site to catch rainwater, then doused the soil and seedlings with compost tea, one of Ingham's favorite ways of ushering microbial life into depleted soil. (First, you have to make good compost, which I later learned in a class with Ingham is a scientifically precise task—she is quite dismissive of the reduced waste or "putrescent slime" that often passes for compost.)

Weatherbee's before and after photos were stunning. A year after he began his work, Mexico suffered the worst drought in 60 years. Fields dried up all around the test site, which stood out like a green flag against a yellow backdrop. At Weatherbee's site, the cornstalks often had three branches bearing ears, versus unbranched, single stalks with solitary ears in the neighbor's field. He suffered no corn borer worms, but they ravaged other fields. His ears of corn were voluptuous things with wild headdresses of red silk. The neighboring farmer—a very good farmer, Weatherbee said—had ears that looked like tiny dry pinecones.

In the afternoon, a lanky Missouri rancher named Greg Judy loped to the front of the room. He declared himself a microbe farmer and called his cattle mobile microbe tanks. He'd been "mob grazing" since 2007, which is a Savory-inspired approach that moves very large herds of cattle over relatively small plots of land—for Judy, that meant about 100,000 pounds of beef per acre. After one year, he had more forage (edible plants for the cattle), more earthworms, and better drought protection—his fields were still lush when his neighbors' went dormant during a drought. He was so successful that he was able to quit his non-farm job. He expanded his operations by leasing nearby plots of land that were so depleted that no one bothered to farm them anymore. He marched his cattle up and down country roads to reach these other plots. They became so lush and filled with wildlife under his management that one hunting-enthusiast neighbor tore up the lease and invited him to mob his cattle through—he hits each property about twice a year—in perpetuity.

Some of my favorite lines from Judy's entertaining spiel:

"Nature did it right for millions of years until we came and boogered things up."

"I don't put hay out for my cattle anymore. You put a bale of hay out for a cow and you turn her into a welfare recipient. My cows work for a living."

"Weeds are what Mother Nature lays down to protect her precious skin."

After Judy spoke, the audience moved outside and mobbed the cookies set out for our afternoon snack. I saw that the man next to me wore a name tag saying that he was from the animal science department at California Polytechnic State University. I was eager to meet more scientists focused on this approach, so I dropped my cookie in my pocket and thrust out my hand. "I guess you and the other scientists here do a lot of studies about this kind of stuff," I said, waving toward Greg Judy.

To my surprise, Rob Rutherford, the animal scientist at the other end of my hand, shook his head. "We do almost nothing," he said. "The farmers and ranchers are way ahead of the scientists on this."

A few months later, I asked the same question of David Zartman, an emeritus professor of animal sciences from Ohio State University. "There's no one to pay for this kind of research," he explained. "A fertilizer company is not going to do it. A pesticide company is not going to do it. The money dictates the direction of the research, not the other way around."

I've always had fond feelings for ag scientists. I grew up not far from the University of California–Davis, which has one of the nation's premier ag departments. When I moved to Ohio, my father used to try to lure me back— he'd call and offer to pay my tuition at Davis *and* buy me a horse. At that point, I was more interested in Mao than moo, but the invitation always clutched at my heart a little. So it was a little disheartening to learn that the corps of ag scientists—concentrated at the nation's 105 land-grant university campuses—wasn't hotly pursuing these exciting new ideas in agriculture.

Really, some of these ideas aren't even so new. Farmers have used cover crops for centuries because they observed that their land was more productive if they grew some sort of plant after they harvested their main crop. They didn't understand the complex interactions between the sun and the atmosphere and the plants and the soil microbes that scientists understand now; they just knew that it made them more successful. So how did such ideas manage to fall by the wayside? How did fertilizing our soils with the kind of chemicals that nearly blew up the town of West, Texas, in the spring of 2013 become the norm?

Blame Abraham Lincoln, although it's likely that he'd be dismayed by the state of American agriculture, too.

Lincoln grew up in rural hardship. As Ricardo Salvador, the director and senior scientist for the Union of Concerned Scientists' Food and Environment Program, told me, "He spent a lot of time on the sharp end of a team of horses, endured drudgery, and really wondered for hours on end whether there was a better way." When he became president, Lincoln set up a research apparatus that would look for that better way. He established the National Academy of Sciences. He formed the Department of Agriculture. He signed a bill proposed by Vermont senator Justin Smith Morrill to set up a network of colleges that would focus on agricultural research and education. President James Buchanan had vetoed the bill—it had also been hotly opposed by representatives from the South—but Lincoln signed it into law in 1862. The law authorized the federal government to deed some lands to each state and territory for these colleges, most of them now universities. Federal and state funds have helped support these land-grant institutions ever since.

In an address delivered in 1931, W.J. Kerr, the president of Oregon State Agricultural College, praised the formation of land-grant colleges for bringing democratized education and an understanding of science to an ignorant country. He said:

> The people generally knew little about science or the applications of science. The common attitude toward applied science was expressed in a New England newspaper of 1816, in opposition to a plan for street lighting, by such statements as these: "Artificial lighting is an attempt to interfere with the divine plan of the world, which called for dark

during the night time." "Emanations of illuminating gas are injurious." "Lighted streets will incline people to remain outdoors, thus leading to increase of ailments by colds." "The fear of darkness will vanish, and drunkenness and depravity increase."

Later in his address, Kerr cataloged the contributions of the land-grant colleges to American agriculture, noting that their work had resulted in the development of whole new industries. All told, he said, the land-grant colleges' contribution to the economic welfare of the country was around $1 billion per year.

Thousands of acres in northern Indiana, formerly unproductive and abandoned by farmers, have been reclaimed for productive agriculture as a result of researches in soils and crops at Purdue. Introduction of new varieties of cane saved the Louisiana sugar cane industry from ruin caused by the mosaic diseases affecting old varieties. Feterita, introduced from Egypt into Texas, has developed into a fodder crop valued at $16,000,000 annually. Grimm alfalfa, brought to this country by a German immigrant whose name it bears, and demonstrated as hardy and adaptable by repeated trials at Minnesota, has replaced the ordinary varieties in the major alfalfa-producing sections of the country. Among new wheat varieties developed by land-grant colleges, such as Kanred, Denton, Federation, and Defiance, Denton alone in Texas has increased the agricultural wealth of the state at the rate of $2,500,000 a year. In Missouri an industry totaling $15,000,000 a year has developed from the introduction by the land-grant college of the soy-bean, a crop that was formerly practically unknown. Prompt control of the cotton boll weevil, the European corn borer, and the Mediterranean fruit fly, saved from ruin the major industries of growing cotton, corn, and citrus fruits. Improved production methods and greatly reduced production costs resulting largely from research, have led to the development of dairy production into a $2,750,000,000 industry. Breeding for egg production in Oregon, the evolution of the trapnest by the Maine station and the day-old-chick enterprise originated at the New Jersey station, have together revolutionized the poultry industry. The peach industry of the Atlantic Coast and Great Lakes regions, and the entire apple industry of the United States, exist today as commercial enterprises because the land-grant colleges have been successful in devising methods of controlling diseases and insect pests.

From the vantage point of 1931— and even looking back from 2013—Lincoln's efforts to boost agriculture and the system that evolved around those efforts have been a massive success. "We licked productivity," Salvador told me. "In Lincoln's time, you really could starve if the crops didn't turn out that year. Now, none of us worries where our food is going to come from. There's never been as much food as there is now, and when we think about it, it's mostly about convenience and what we're going to have. Even those who worry about food don't really worry about food. They worry about whether they have the money to buy the food. The food is not the question."

But tucked into all this energetic tinkering with agriculture was the dangerous idea that humans had a mission to conquer nature for their needs. It's an idea that rose from the Enlightenment, says Fred Kirschenmann, the former director and now a distinguished fellow at the Leopold Center for Sustainable Agriculture at Iowa State University.

"The Enlightenment freed us from what we call the Dark Ages and from the ideologies that had locked us in," Kirschenmann told me. "But a part of that Enlightenment vision was that we humans were somehow separate from nature, that we could not only dominate it but had a responsibility to do so. René Descartes once famously said that we had become the masters and possessors of nature. There was an assumption that nature was this collection of materials that we could manipulate for our sole benefit. There was no awareness that nature actually was a living community that's very interdependent."

The catastrophic apex of this kind of thinking thundered across the High Plains just a few years after Kerr's remarks. Sodbusters eager for big profits growing wheat ripped up virgin prairie soil and planted their crops over and over again, without giving anything back to the land—no cover crops, no mulch, just relentless plowing. Without vegetation to nourish the soil and hold it in place, the land baked and bleached and dried. When the onslaught of plows coincided with a long hot spell and high winds, the Dust Bowl was born.

Timothy Egan captured the Dust Bowl masterfully in his book *The Worst Hard Time*. "Dust clouds boiled up, ten thousand feet or more in the sky, and rolled like moving mountains—a force of their own," Egan writes. "When the dust fell, it penetrated everything: hair, nose, throat, kitchen, bedroom, well. A scoop shovel was needed just to clean the house in the morning. The eeriest thing was the darkness. People tied themselves to ropes before going to a barn just a few hundred feet away, like a walk in space, tethered to the life support center."

When the dust mountains rumbled to the east with a storm in May 1934, the rest of the country had to take notice of this environmental disaster on the Great Plains. "In Chicago, twelve million tons of dust fell," Egan reports. "New York, Washington—even ships at sea, 300 miles off the Atlantic coast—were blanketed in brown."

Mother Nature made her point about the perils of ruining the soil, and we took heed, at least for a while. Just a few weeks after a second cloud of Great Plains dust blackened the midday skies in Washington, DC, a rattled Congress passed the Soil Conservation Act of 1935. The law recognized that "the wastage of soil and moisture resources on farm, grazing, and forest lands . . . is a menace to the national welfare," and it directed the secretary of agriculture to establish the Soil Conservation Service as a permanent agency—it is now known as the Natural Resources Conservation Service. Two years later, President Roosevelt authorized the creation of soil conservation districts throughout the country, and 3,000 of them operate today. Egan believes that the work done by these districts changed practices in the Great Plains such that despite two more significant droughts, Dust Bowl II never arrived.

Not to say that the dust storms have stopped completely. The first time I googled "dust storm" a few years ago, a citation popped up for one that had just happened 2 weeks before. I just googled it again, right now, and found a January 2013 article about dust from east-central Colorado that blew into northwest Kansas and reduced visibility to a quarter mile, causing hazardous conditions for an hour.

In addition to changes in public policy, the Dust Bowl prompted a sort of agrarian renaissance in which people rethought not only the way they treated the land, but also their overall relationship to it. "Do civilizations fail because soils fail or do soils fail because civilizations don't know how to take care of the ground beneath their feet?" That sounds like something Elaine Ingham or Gabe Brown or Allan Savory might say, but it's a quote from the USDA's 1938 *Soils and Men: A Yearbook in Agriculture.*

One of the most prominent suitors of the soil was Louis

Bromfield, a Pulitzer Prize–winning author and Hollywood hobnobber. He returned to Pleasant Valley in Ohio after 13 years in France to spend the last part of his life restoring the soil on eroded land near the family farm where he grew up. He wrote two books of short stories and five novels in Ohio, but none were critically acclaimed—in fact, Edmund Wilson titled a fierce review of one of his books "What Became of Louis Bromfield" in the *New Yorker*. But at this point, Bromfield was only writing fiction to support his work on the farm. When he started writing nonfiction about that experience, the critics once again applauded.

Bromfield couldn't benefit from the kind of soil science available to today's innovative farmers, as studying soil health and belowground biology is a fairly recent phenomenon—in fact, Agricultural Research Service (ARS) soil scientist Doug Karlen told me that when he offered to do a workshop on soil health at the Soil Science Society of America's annual convention 15 years ago, some members of the society sneered that soil health wasn't science. But Bromfield was a careful observer and tried to work with natural processes instead of trying to circumvent or battle them, as a growing number of "modern" agricultural experts were urging. He initiated careful cropping and grazing strategies, with splendid results. Abe Collins is something of a scholar of the agrarian renaissance of the 1930s and 1940s; his e-mails trail a long list of quotations from soil-building pioneers like Bromfield. "Experts say it takes nature 1,000 years to make an inch of soil," Collins told me. "But Louie was building one to two inches of soil in a year and changing the soil profile at depth. He bought worn-out farms and had a foot of really fertile soil in ten years."

My last foray into the Ohio countryside before I moved to Portland was to Bromfield's Malabar Farm, now a state park just a few miles from the interstate. Corn was lined up like thousands of soldiers in battle formation on both sides of the road, cows were eating hay in a muddy paddock—the place didn't scream enlightened farming. Aside from his books, which continue to inspire people like Abe Collins, Bromfield's only enduring legacy is the 32-room house he built for his family and guests, and his huge collection of books and art. I joined a crowd of visitors with fidgety, unimpressed children for the house tour, marveling with a trio of librarians over the Grandma Moses paintings and the framed *New Yorker* cartoons that referred to Bromfield. I gaped at the staircase where Lauren Bacall tossed down her bridal bouquet to the gang of friends who went rustic for her wedding to Humphrey Bogart. And I coveted Bromfield's amazing bedroom, with his huge writing desk and a bed built into a bookcase and walls painted green and red. I was sure that the illustrator for *Goodnight Moon* must have spent some time there.

Fortunately, another urban defector from the 1940s had a more lasting and profound effect on America's appreciation for the soil. J.I. Rodale—born Jerome Irving Cohen—started life in the Lower East Side of New York, where he was the sickly son of a grocer. He moved through a number of careers before he encountered Sir Albert Howard's book, *An Agricultural Testament,* in which Howard describes his years as an agricultural expert in India from 1905 to 1931, where he discovered more wisdom in the practices of local farmers than in the practices he had brought from England. Howard became a champion of building soil health and developed a method of composting waste materials into powerful soil amendments. Rodale bought a farm to practice Howard's methods and became the first person to popularize the organic approach in America. When Rodale launched his first magazine in 1942, *Organic Farming and Gardening,* Howard was an associate editor.

Rodale sent out 10,000 free copies of his first issue to farmers, but they weren't a receptive audience back then; not one subscribed to the magazine. Thirty years later, an article about Rodale titled "Guru of the Organic Food Cult" in the *New York Times Magazine* reflected that "chemical fertilizers are considerably easier to apply than bulky organic matter and tend to provide larger yields during routine applications. J.I. was spreading his gospel at the start of a full gallop in the American farmer's movement toward chemical fertilizers; their use in the U.S. has increased sevenfold since 1940."

But Rodale persevered. By the time the *Times* article came out, he and his work were hugely influential among an array of subcultures. The *Times* writer tagged them as "food cultists from old-line vegetarians to youthful Orient-inspired 'macrobiotic' dieters with their emphasis on whole grains, especially rice, plus reactionaries yearning to turn back all clocks, urban dropouts in search of simpler, more natural life styles, ecologists who are worried about the long-range environmental effects of some chemicals, Dr. Strangelove paranoids who read poison plots on the ingredient labels of pancakes mixes and increasingly, rather ordinary folk to whom pronouncements about the perils of cyclamates, phosphates, etc., have stirred a wariness about all man-made chemicals, particularly those that get into their food, or that they think do."

I number myself among the Drs. Strangelove.

The *Times* article portrays J.I. as a somewhat endearing crackpot, but, really, so many of his concerns were incredibly prescient. As proof of Rodale's kookiness, the *Times* writer points out that he considered wheat, sugar, and fluoride in drinking water to be unhealthy, but those are quite mainstream concerns now. When the writer asked why—if organic agriculture was so superior—it hadn't been more widely adopted, J.I. complained that the entire agricultural establishment, from the land-grant universities and academic agriculture to the Department of Agriculture, was held in thrall by the profit agenda of agribusiness. Plenty of people share that view today! I heard a similar twist on that argument from a prominent physician a few years ago when I was writing an article about gluten-free diets. He told me that celiac disease, in which the gluten in wheat and other grains actually damages the digestive architecture of affected people and can cause precancerous conditions, was hugely under diagnosed in the United States compared to other nations. Why? I asked. He thought that the disease attracted few research dollars because Big Pharma didn't see a money pool there, since the best treatment wasn't drugs but simply to avoid glutenous foods.

J.I. turned his attention to writing plays in his later years, and his son Robert took over the family enterprise. Like his father, he bent ears hither and yon about the superiority of agriculture that worked with nature; as they had with his father, many people shrugged and turned away. On a trip to Washington, DC, in the late 1970s, he met with lawmakers to try to win public policy considerations for organic agriculture. The response was "Come back when you have the research to back your claims that this is a better or even credible approach." So Robert Rodale headed back to the Rodale Institute in Kutztown, Pennsylvania, and, with an agronomist named Richard Harwood, designed a side-by-side trial of organic versus conventional farming. The trial continues today and is the oldest such experiment in the United States and the second oldest in the world.

The Farming Systems Trial (FST) marked its 30th anniversary in 2011. The following spring, I drove from Cleveland to Kutztown to meet with Jeff Moyer, who's worked on the trial from the very beginning and is now the Rodale Institute's farm manager. We clomped from his office toward the test fields, passing several cylindrical compost piles at the heart of the campus that steamed like giant cups of coffee—as the microbes eat, work, and reproduce, they build up a lot of heat. The test fields themselves were pretty unremarkable looking—nothing much was growing—but it was clear that neither the organic nor the conventional plots had any sort of topographic advantage, as they alternated in 5-foot strips for 100 feet across the landscape.

The organic strips employ the basic three tools of organic agriculture—the "three Cs," as Moyer called them: a top dressing of good compost, nitrogen-grabbing legume cover crops during fallow periods, and crop rotation. The thinking behind the latter is that when you plant the same crop in the same field every year, the diseases and pests that prey on that crop take up residence there, too; if you move the crop every year, then you don't have to fight so much with the diseases and pests. Good organic farming isn't just the absence of synthetic chemicals, Moyer explained, but rather an approach that works with biological processes and regards the soil as a complex system of living organisms. "We don't talk about soil quality anymore," he said. "We talk about soil health. It seems like a game of semantics, but I think it's important. I can have a high-quality table, for instance, but I can't have a healthy table."

The FST has involved many partnerships with USDA and other scientists, and inspired many master's theses and dissertations—Moyer guesses the number as 40. The trial only compares corn and soybeans, since about half the cropland in the United States involves those two plants and that's where the benefits of organic could have a huge impact on our land, water, and rural livelihoods. Over the years, the Rodale Institute has changed certain aspects of the trial so that it fairly represents the differences between modern organic and conventional agriculture. For instance, conservation requirements in the 1985 farm bill pushed many otherwise conventional farmers to go no-till, so the FST split its conventional plots into till and no-till to account for that change. And since 94 percent of all soybeans and 72 percent of all corn grown in the United States are now genetically modified to either resist the effects of herbicide or to produce their own pesticides, the FST incorporated those varieties into the conventional plots.

Looking back over 30 years, the results are a resounding endorsement for organic. The organic plots built soil carbon, while the conventional plots depleted it. After the first 3 years of transition, the organic plots and the conventional plots produced the same amounts of food, except during drought years—and then, the organic yields were 31 percent higher. The organic corn and soybeans survived weedy encroachments better than the conventional crops—a huge issue, since the herbicide-resistant genetically modified crops are so heavily sprayed with Roundup and other glyphosate-based weed killers that the practice is creating swaths of superweeds that are resistant to herbicides. At last count, there were 197 such superweeds. And finally, the organic systems were nearly three times as profitable as the conventional systems, because they are produced without costly chemicals and consumers are willing to pay more for them.

Scoffers would scoff, saying things like "Sure, what do you expect Rodale's tests to prove?"—despite the fact that Rodale has partnered with many outside scientists on the trials, including Matt Ryan from Cornell University, Michelle Wander from the University of Illinois, and David Douds from the USDA's lab in Philadelphia. But other organizations have conducted studies that come to the same conclusions. A report from the United Nations noted that "Organic agriculture has the potential to secure a global food supply, just as conventional agriculture is today, but with reduced environmental impact." A study from Iowa also found that organic systems for corn and soybeans produced about as much food as conventional systems over 12 years. A study from the University of Minnesota showed that farmers who use genetically modified crops made less money over a 14-year period—the seeds and chemicals that they're engineered to work with are expensive and make profit margins wither.

All of this raises questions: Why don't we know about any of this stuff? Why is it that people who set out to practice high-knowledge, low-technology, soil-cherishing agriculture are dismissed as either nostalgic Luddites or elitists who only care about feeding people who can pay a premium for boutique food? Why is it that even people who shudder at the damage conventional agriculture does to the environment conclude that this kind of agriculture is a necessary evil if we're going to feed 9.6 billion people by 2050?

"People frame us as being anti-modernity or anti-progress because they equate the new technologies with progress," Ricardo Salvador told me. "And when we talk about cover crops and integrated pest management and poly-cultural systems and crop rotation, that sounds suspiciously like the kind of agriculture that was practiced in the past. But there's a sophistication and real progress that comes from taking what we know about biological systems and applying it. We could be far more efficient in our agriculture if we only applied this approach, but of course there's no profit other than to the farmers themselves, and those that would benefit directly when the farmers benefit."

That would be us, the consumers.

I went to a conference of food and science writers a few years ago where an array of speakers held forth on an array of food-related topics. The title of one speaker's talk was "The Culture War in Food and Farming: Who Is Winning?" I listened with interest at first, then with growing umbrage. Who favored organic agriculture? the speaker asked, and then answered his own question. Not economists or scientists or the public, he said. Rather, it was the cultural elite like Michael Pollan and Barbara Kingsolver.

That didn't sound right. By this time, I had spoken to many scientists who were interested in alternative agriculture and who talked about how hard it was to get funding for anything other than projects that fall within conventional agriculture's bailiwick. I knew plenty of ordinary people who wanted more wholesome food; I was delighted when I'd see shoppers at my farmers' market buying organic produce using public assistance cards. I thought that was a damned good use of my tax dollars. So I Googled the guy while he was talking and found that he had served as a consultant to the leadership of Monsanto. I raised my hand, stood, and said all of that. He approached me after he finished speaking to tell me that he was never paid by Monsanto. Was it better that he had worked for them for free?

The sad fact is that the system that Lincoln set up to create better agriculture now largely serves agribusiness, instead of the individual and everyman farmers and the nation of eaters whom he wanted to help. And it's the agribusiness narrative that's upheld, no matter how much proof emerges that there is another approach to agriculture that will not only stop destroying our environment, but also heal it. Not only does Big Ag keep telling us that we need to adhere to their agenda, they keep telling us we need more of it.

"We still operate out of the cultural meme that we humans are superior to everything else in nature, that we're the big-brain mammals and that we can always come up with the technology that's going to be superior to anything nature does," Fred Kirschenmann told me. "That's really behind the question that is posed so often in the popular press: 'How are we going to feed 9 billion people?' The assumption is that we just have to push the pedal to the metal a little harder and figure out the new technologies that make that happen."

The specter of 9 billion gaping, hungry mouths is worrisome, of course—there are already almost a billion people struggling to get enough to eat, and who wants to see that number grow? But careful observers say this image is actually just a club crafted and wielded by agribusiness to scare us into supporting their agenda. We already grow more than enough food for 9 billion people, according to calculations done by the Millennium Institute. Global agriculture currently produces 4,600 calories for every man, woman, and child on earth—twice as much as we need to thrive. The problem is not one of production, but of distribution.

"The loss of food from retail on is huge," said Hans Herren, an agronomist who is president and CEO of the Millennium Institute and was one of the editors of the UN's global report "Agriculture at a Crossroads." "In the developed world, people throw away more than 30 percent of their food after they buy it. The price is so low that they don't value it."

The narrative from the Farming Systems Trial and from innovative farmers who have largely left chemical, conventional agriculture behind challenges powerful vested interests. Even individual farmers caught up in conventional agriculture resist the narrative from regenerative agriculture because they have made such huge investments in the current system. Back in the 1970s, Earl Butz, head of the USDA under Richard Nixon and Gerald Ford, roamed the country urging farmers to take on debt so that they could grow their businesses and produce for the global market. "Plant fencerow to fencerow" and "Get big or get out!" were two signature Butz bellows. So farmers followed the advice the USDA and the academicians pressed on them, plowing up more land and buying expensive machinery—when you see a John Deere combine churning down a field, that alone is a nearly half-million dollar investment—and charging after short-term profits.

"When you put yourself in that position, what are you going to do?" Fred Kirschenmann told me. "You're going to do everything you can to keep the current system going, because you're over 60 now and that's what you've invested in."

Laura Jackson, an evolutionary biologist at the University of Northern Iowa, describes the agriculture that has resulted from the Butz revolution as a giant open-air factory owned by monopolies and stretching across the Midwest. You can drive for 6 hours and see nothing but corn and soybeans, and before and after these annual crops are planted and harvested—a period of about 9 months—the ground is naked and exposed to the elements. This is agriculture, but we don't eat what it produces. The corn and soybeans are commodities that feed cattle and other animals in feedlots, supply manufacturers making high fructose corn syrup and other food additives, and provide the raw material to ethanol makers.

Jackson suggests that the allegiance on the part of conventional farmers to this kind of agriculture is an example of Stockholm syndrome. "If you become a captive and begin to look at your captors as friends and allies instead of what they are—kidnappers—then you suffer from Stockholm syndrome," she said in a speech. "In the pressure to comply with the demands of your kidnappers, you forget who you are, who your real friends are, and who your family is. . . . Our high-energy corn and bean system is a kidnapper, not our friend, even though we may have to get along with it for the time being because it's all we have."

Thirty years of concentration and consolidation in agribusiness—in meatpacking, seeds and chemicals, grain handling and shipping, farm equipment, fertilizers, and food retailing—has allowed these industries to winnow down the opportunities for farmers to just a few. The seed industry is a good example. Neil Harl, an emeritus professor of agriculture and economics at Iowa State University, wrote a brief to the Department of Justice recommending anti-trust action against the seed and agricultural chemical industry, noting that farmers could choose from among 300 suppliers of corn seed in the 1970s. The competition among these suppliers kept seed prices affordable.

This changed with the advent of agricultural biotechnology and laboratory-manipulated genetic material. The United States Patent Office had never favored the patenting of life forms and turned down a patent for a microorganism that ate oil spills. But the US Court of Customs and Patent Appeals overturned the case, and the Supreme Court upheld that decision in 1980, ruling in *Diamond v. Chakrabarty* that a life-form could be patented. A little

more than 20 years later, the Supreme Court ruled that
seeds, too, could be patented. For the first time, farmers
were prohibited from saving seed from these patented biotech crops to replant the following year.

The technology to create these new bio-tinkered seeds
is expensive, and the number of seed suppliers dwindled
rapidly to the handful of big firms that could afford it. Now,
Monsanto controls 30 to 40 percent of the market directly,
creating seeds that—in the case of Roundup-ready crops
that resist herbicides—are designed to coexist with large
amounts of Monsanto chemicals to perform as advertised.
Harl estimates that Monsanto actually influences pricing
for 90 percent of corn and soybeans seeds, since they also
sell biotech traits to competitors. Given Monsanto's control
of the market, it is not surprising that the price of seeds for
farmers has skyrocketed: Prices rose 150 percent between
1999 and 2010. Some farmers are able to absorb the higher
costs, but many marginal producers have been driven out of
business.

Harl is a tenacious advocate for antitrust action
against agribusiness, but he's not optimistic that without
pressure from consumers, the government will go after
these huge monopolies that control our food. "There's a
huge amount of money and a lot of pressure applied whenever someone in Washington tries to do something about
this," he told me. "That pressure is applied in the form of
messages like 'Look, if you let this go on, we're going to
diminish our support for your campaign.' When things get

bad enough that consumers rise up, that's when we'll get another era of antitrust."

How much money is involved? The Center for Responsive Politics reports that during the past 2 decades, the agriculture sector contributed $480.5 million to national political campaigns. In 2009, the sector spent $133 million on lobbying. This was nearly as much as the nation's defense contractors spent that year, but way below the $385.9 million that the energy sector forked over in 2008.

The government not only won't go after Big Ag, but also uses our tax dollars to support agribusiness and the system of ruinous agriculture it has created. "It's the most perverse way you could set up a civilization, where you're paying the farmers to degrade their underlying resource," Abe Collins told me in one of our early conversations. He meant that through the farm bill, which has been redrafted in various ways since the Great Depression and has been revised about every 5 years since 1973, we as taxpayers reward farmers for practicing the kind of agriculture that depletes soil—at a cost of $95 billion annually. It took me a while to understand the mechanism for this, since US agriculture policy is so complex—really, I wanted to find a "Farm Bill for Complete Idiots" book to break it down for me. Fortunately, I discovered the Web site of Harvest Public Media, which reports on "food, fuel, and field," where an article by Kathleen Masterson offered a lucid explanation of the history and current state of government payments to farmers.

Until 1996, the government gave crop subsidies to the nation's farmers. The practice began during the Great Depression to help farmers stay in business, because their income fell off precipitously whenever food prices dropped due to overproduction. The subsidies paid farmers to leave some land fallow to avoid glut. The government also bought excess grain, which it stored and later released during times of need.

Grain prices were high in the mid-1990s, prompting the architects of the 1996 farm bill to announce provisions that would reduce and eventually eliminate subsidies. But when prices fell once again, new government programs were quickly crafted to keep farmers in business. One program sent farmers direct payments based on the number of acres they had farmed in the 1980s, regardless of market conditions. Another automatically sent farmers a check when prices dropped. The payout soared even higher. American farmers received more than they ever had: $20 billion annually from 1999 to 2001.

The consequence of these government programs was that farmers planted more crops, since the more acres they farmed, the higher their government check would be. "I think the net effect was farm consolidation, concentration and a decline in opportunities for new farmers to get started," Ferd Hoefner of the National Sustainable Agriculture Coalition told Masterson. "There were lots of reasons why farms got bigger and fewer in number and certainly technology played a big role, but policy certainly has contributed to that."

As is so often the case, these government programs that were supposedly intended to help *all* farmers disproportionately helped the very richest ones—those with vast acreages who produced the five major commodity crops of corn, soybeans, wheat, cotton, and rice. From Masterson's article: "According to the Environmental Working Group's crunching of U.S. Department of Agriculture numbers, between 1995 and 2010 three-quarters of farm subsidy dollars go to the top 10 percent of those who receive subsidies. About 62 percent of American farmers don't receive any subsidies at all, according to 2007 data." Many of these farmers are raising fruits, vegetables, and nuts that aren't covered by subsidies. Ironically, they're producing food that we actually eat as opposed to the crops that enter the juggernaut of agricultural processing and emerge, for the most part, as nonedible products. Even some farmers producing commodity crops don't get subsidies because they're working acres that weren't mapped when the subsidy programs were set up.

The 1996 farm bill contained two other major changes. It required farmers to purchase crop insurance that was heavily subsidized by the government to be eligible for farm payments, and it ended the requirement that they engage in conservation practices in order to be insured. Enrollment in crop insurance was swift and heavy, and farmers received nearly as much federal assistance from insurance payoffs as from the direct payments. Total payments to farmers in 2012 were $20.45 billion.

High as it is, that figure doesn't reveal the full flow of taxpayer dollars underwriting industrial agriculture.

According to a 2004 study by agricultural economists Michael Duffy and Erin Tegtmeier at Iowa State University, the government was spending another $5 to $16 billion per year to mitigate some of its damage to our environment; the costs are likely to be significantly higher now. "Industrial agriculture only stands up economically because we don't take into account the damage to our soils, the way we're exhausting fossil fuels and our mineral reserves, and the depletion of our groundwater," Ricardo Salvador told me.

Given the economic crisis that began in 2008, the fact that the government funnels tax dollars to so many farmers who already have comfortable incomes seems scandalous. Still, federal crop insurance continues to do everything to support conventional agriculture and very little to support alternatives. Farmers pay a reduced rate for crop insurance when they plant seeds that are Roundup-ready or have some other genetic modification, because these crops are seen as being less risky than non-GMO crops. On the other hand, they pay a premium on their crop insurance if they farm organically. And farmers who plant cover crops—cover crops!—can actually be denied crop insurance in some situations. Farmers like Gabe Brown who plant companion crops—clover along with his oats, for instance—are booted from the program entirely. "I'm saving fossil fuels and fertilizer and improving soil health, but I'm penalized for it," Brown told me. "The government knows nothing about how nature works."

"You'd probably see a lot of changes in the system if

we stopped crop insurance," explained Cornelia Flora, an emeritus sociologist at Iowa State University. "It's true that farming is a risky business, but the insurance lets you do anything you want. If you're insured against crop failure and you're insured against price failure, why would you ever do anything different from the way you've always done it— because that's when you *won't* be able to get crop insurance. In other situations, a risk reduction strategy might be crop diversification or something more systemic than just saying the government will bail me out."

The land-grant universities that Lincoln so proudly signed into being have also been compromised by a huge influx of corporate cash. These public institutions include some of our largest and most prestigious centers of learning, including the University of California system, Pennsylvania State University, and Texas A&M University. Beginning in the 1970s (and aside from a brief resurgence in the 1980s), federal and state support for these institutions—covering both their research and operating expenses—began to dwindle. Agribusiness has eagerly filled the gap.

One particular financial blow to the land-grant universities has had a direct impact on farmers and ranchers. The USDA has steadily reduced funding to the agricultural extension services that used to pump research findings from the universities to agriculturalists via training and educational materials. Farmers used to bring their problems to these extension agents, who would offer insights from the latest research. But with the number of extension agents dwindling, farmers have few options but to turn to the agronomists and those stationed at the grain elevators, where they buy seed and chemicals.

"Back in the 1980s, the extension agents trained people at the elevator to be their eyes on the ground," said LaVon Griffieon, an Iowa farmer who raises antibiotic- and hormone-free beef, pork, chicken, turkey, and lamb. "But all they have is a magic potion for your problem. It's all solved chemically, and we don't get that extension of knowledge from our land-grant universities that we used to get."

But a damning report issued by Food and Water Watch in 2012 indicates how shaky that knowledge itself has become. "By the early 1990s, industry funding surpassed USDA funding of agricultural research at land-grant universities. In 2009, corporations, trade associations and foundations invested $822 million in agricultural research at land-grant schools, compared to only $645 million from the USDA (in inflation-adjusted 2010 dollars). . . . Industry-sponsored research effectively converts land-grant universities into corporate contractors, diverting their research capacity away from projects that serve the public good."

With less federal funding for research, land-grant university faculty under pressure to conduct research and publish the results—this is what determines tenure and salaries—have to go where the funding dollars are flowing, which is the private sector. Of course, they can't necessarily get funding for the work that they're really interested in; their research has to reflect a corporate agenda to win corporate dollars. That corporate funder then benefits from the tax dollars underpinning that scholar, student body, and university. Robert Taylor, an economist at Auburn University who studies agribusiness structure and concentration, told me, "Let's say I get a $100,000 corporate grant for field experiments. I still have most of my salary paid from taxpayer funds, but I devote all my energies to the $100,000 project. This leverages taxpayer funds in a fuzzy way and, in effect, becomes a corporate subsidy."

Nearly half of all land-grant ag scientists surveyed in 2005 reported that they had received research funds from the private sector. One of the problems with this is the so-called funder effect: Studies show that industry-funded research is more likely to reach favorable conclusions for that industry. And often, the results of these studies are published in journals that don't require the study's authors to disclose the sources of their funding. Still, this research often guides policy makers and regulators.

When I talked to Tim Schwab, the lead researcher for the Food and Water Watch report, he said that both the USDA's research arm—the Agricultural Research Service— and the land-grant universities are failing to conduct the independent agricultural research we need. "When you look for research on the safety or environmental impacts of genetically engineered crops, there just isn't much coming out of our universities," he said. "The ARS should be stepping up to the plate and doing the really robust necessary research, but they're not."

Schwab pointed to a recent kerfuffle to make his point. French molecular biologist Gilles-Eric Séralini published a 2012 study of Monsanto's genetically engineered corn and the Roundup herbicide used with it and found major health concerns. Detractors and even science journalists said that Séralini was ideologically driven and had been funded by a foundation that was biased against genetic engineering (and one critic of industrial agriculture even told me that Séralini used the mouse strain and sample size to get the results he wanted). Schwab told me that science journalists routinely fail to notice the converse: that industry has funded most of the studies showing that genetically engineered seeds are safe. "The kind of corn Séralini studied has been shown in the scientific literature to be safe and effective," Schwab said. "But all the studies I could find on it were performed by or paid for by Monsanto."

As the Food and Water Watch report shows, corporate influence at the land-grant universities goes well beyond funding research. Donations from agribusiness announce their brands across campus: Witness the $1 million donation that resulted in the Monsanto Student Services Wing at Iowa State University, the $200,000 donation to the University of Illinois that resulted in the Monsanto Multimedia Studio, and the named ConAgra and Kroger research laboratories at Purdue. Some universities sell seats on academic research boards to corporations. For instance, a $20,000 donation gets industry sponsors seats on the University of Georgia's Center for Food Safety's board of advisors, where they can influence research direction. Takers include Cargill, ConAgra, General Mills, Unilever, McDonald's, and Coke. Industry also spends millions endowing faculty chairs. "This list could be so huge it would be comical," the Union of Concerned Scientists' Ricardo Salvador told me. "The most notorious is the case of Novartis literally buying the entire College of Natural Resources at Berkeley.*"

The fusion of corporate and university interests is also stunning at South Dakota State University (SDSU). University president David Chicoine joined Monsanto's board of directors in 2009 and received $390,000 the first year. At around the same time, the university joined the Monsanto subsidiary WestBred in its effort to sue farmers for seed-patent infringement—a stark contrast to the old days, when land-grant universities developed public seeds which farmers could use, save, and share as they wished. "What makes the university lawsuits against farmers more offensive is the fact that SDSU wheat seeds were developed with farmer and taxpayer dollars," says the Food and Water Watch report.

Unless researchers win grants from the dwindling pool of federal dollars doled out by the USDA, the National Science Foundation, and the Department of Energy, they're unlikely to be able to conduct research on alternative agriculture or investigate problems with conventional agriculture.

---

*This was a 5-year partnership that was not renewed.

Even the USDA's research agenda is biased toward conventional agriculture—that's the system that's in place, and most of their funding is based on improving that system, not changing it. And even if researchers win federal dollars to conduct experiments in regenerative agriculture or investigate problems with Big Ag, they risk antagonizing university administrators, who keep an anxious eye on the continuing good graces of their corporate sponsors.

"It's not that you have companies like Monsanto controlling the land-grant universities, but they're always in the shadows," Fred Kirschenmann told me. "Administrators are always looking over their shoulders."

Kirschenmann points to the career of one of his young colleagues at Iowa State University as an example of the subtle chilling effect this corporate pressure can exert. Ecologist Matt Liebmann had an experiment going for 9 years in which he compared the conventional two-crop rotation of corn and soybeans with three- and four-crop rotations. With these more complex rotations and more diverse crops, Liebmann found that farmers could reduce their fertilizer use by 90 percent and pesticide use by almost 90 percent, with comparable yields and more money going to the farmer. "The university has its own public relations system and they select what they're going to publicize, and they never even mentioned Matt's work," Kirschenmann said. "It wasn't until Mark Bittman found out about it [through the Union of Concerned Scientists] and did an article in the *New York Times* that the public found out."

Liebmann conducted another study looking at the habits of two varieties of field mouse, the prairie deer mouse and the white-footed mouse. He found that if farmers don't till their fields in the fall to make planting quicker in the spring, those two breeds would eat up to 70 percent of the weed seeds in the field and dramatically reduce the need for herbicides. Again, the university PR system ignored Liebmann's work.

Kirschenmann's own story is an example of the less-subtle chill—more like an icing—triggered when someone at a land-grant university seems as if they're issuing too much of a challenge to industrial agriculture. Kirschenmann is a North Dakota philosopher with a 2,600–acre organic farm and an international leader in sustainable agriculture who in 2000 was hired as the director of the Leopold Center for Sustainable Agriculture at Iowa State University. Shortly after his appointment, he called together a meeting of leaders in sustainable agriculture from outside the university, invited them to read the state legislation that created the center in 1987, and asked for suggestions in fulfilling its mission. Karl Stauber, President Clinton's undersecretary of agriculture for research, education, and economics, spoke up. "It's clear that the Leopold Center is supposed to be a center for change," Kirschenmann recalls him saying. "If you're going to be a center for change, then you have to recognize that the people who are in power don't want change. The people who are going to be interested in change are the people who live on the fringes."

Kirschenmann embraced this. At the vibrant fringe, he

found groups like the Practical Farmers of Iowa, a group of individuals younger than the average farmer (58) who were conducting on-site research to help themselves and others create a better agriculture. He now thinks he may not have paid enough attention to the mainstream agricultural entities. "I'm not a skilled politician," he admitted to me.

One of the things Kirschenmann did that seemed to rankle was organize a meeting of smaller pork producers—not farmers with hundreds or thousands of pigs in CAFOs (concentrated animal feeding operations) that result in undifferentiated commodities, but family-size operators—along with university faculty and food-industry reps to talk about niche marketing of differentiated pork products. Thirty farmers were interested, and Kirschenmann helped them form the Leopold Center–supported Pork Niche Market Working Group.

Kirschenmann called the Iowa Pork Producers—the group representing commodity CAFO operators—and asked to speak to their board at their next meeting to discuss new opportunities for smaller producers. He spoke for 15 minutes. When he finished, the president of the board began to shout and pound his fists on the table. "You're just like everyone else who's trying to make us look bad!" Kirschenmann recalls him saying.

Kirschenmann thinks that these big players in industrial agriculture became more and more dissatisfied with his leadership of the Leopold Center and pressured the university to make a change. In 2005, 5 years after he arrived, he was relieved of his position.

Still, Kirschenmann is hopeful. Despite the early hostility that precipitated the loss of his job at the Leopold Center, Iowa Pork Producers and the Pork Niche Market Working Group have worked together cooperatively for the past decade or more. And in addition to his enduring Iowa connection, he's involved with the Stone Barns Center for Food and Agriculture in New York, and every year they harness some 15 to 20 apprentices to the work of growing food sustainably. The apprentices often move on to work on independent farms and then launch their own businesses. Kirschenmann is watching the creation of a new wave of farmers, who are headed out to lap the land with a new kind of husbandry that embraces the complexity of Mother Earth.

Back in the early 1970s, my beloved father- and mother-in-law were selling their family business. They lived in a small town in the Catskills and the buyers were from the big city, and at one point, my father-in-law thought the buyers were trying to take advantage of them. "Do they think we're *farmers?*" he sputtered to my then-husband and me.

I've often turned that over in my head as I've walked through a farmers' market. My father-in-law used the word "farmers" to imply stupidity—and sure, Iowa farmer LaVon Griffieon did tell me that applying Roundup is so easy you could send a monkey out to do it. But many of the young farmers tending their stands at the market know the science of turning sunlight into asparagus or strip steaks, and they enjoy holding forth about it to the crowds of shoppers. They're like rock stars. Vert, very smart ones.

CHAPTER 7

NEW BEDFELLOWS

It was dusk when I arrived at Davis for the California Rangeland Conservation Coalition conference in February of 2012, and there were still enough slants of sunlight to see orange balls dangling from trees along the city streets. Nice, I thought; a lingering civic adornment for Christmas. Then I realized they were actual oranges. I had somehow forgotten what late winter is like in this part of California, just 60 miles from where I grew up.

The conference was billed as an event for ranchers, range scientists, environmentalists, and others interested in protecting rangeland—which occupies about half of California—and keeping it healthy. Here, I felt positively naked without a cowboy hat, which was so much the *chapeau du jour* that I feared all sorts of people were going to start addressing me as "little lady." (I'll generalize that there is a handsome graciousness on the part of cattle people, whether they're men or women.)

The conference organizers took pains to mix the crowd and assigned seats for every meal. Thus, on day two, I found myself sitting at a table with, among others, a scientist from the university and a couple of grad students, a few people from public agencies, a woman from Defenders of Wildlife on my left, and a rancher from Kansas who had just given a talk about his own conservation work on my right. For a while I just listened to the conversations going on around the table, then I asked Jessica Musengezi from Defenders of Wildlife and Bill Sproul, the rancher, if they would have ever imagined themselves sitting together at lunch 5 or 10 years ago.

She smiled and shook her head.

Sproul regarded me solemnly. "If we had been sitting at the same table 10 years ago," he said, "we would have been on opposite ends. And we'd have been sitting there because one of us was suing the other."

I wasn't that surprised. At the Quivira conference the previous fall, I'd not only been told that academic agricultural science didn't take much of an interest in the soil health movement, I'd also been told that many environmentalists weren't attuned to this movement with the kind of wild enthusiasm I'd expected.

One of the speakers at Quivira was Sara Scherr, president and CEO of EcoAgriculture Partners, an organization that advocates for land use that provides rural livelihoods, ecosystem services, and productive agriculture. She spoke of the wariness some climate activists express about the idea of soil carbon sequestration, and later explained the divide over the course of some e-mails and a phone call.

Since the early years of the UN Framework Convention on Climate Change in 1992, the negotiators—who are principally meteorologists and energy specialists—were biased against climate change mitigation strategies that focused on land use, Scherr said. Their focus was on reducing fossil-fuel emissions and transforming the energy sector, and they didn't want to dilute that focus—even though greenhouse gas emissions from land use accounts for some 30 percent of total emissions—and they didn't understand or trust that agriculture could actually remove carbon from the air and sequester it in the soil.

"Their perception was that land use was too variable and complex to deal with, relative to the 'simplicity' of energy, in the eyes of energy experts," Scherr told me.

Moreover, these environmentalists didn't trust the idea of selling carbon credits based on soil sequestration. They didn't believe that the carbon could be stored there permanently and even if it could, they didn't believe that doing so wouldn't cause "leakage"—more carbon-dioxide producing activities elsewhere. "They thought it would be a boondoggle and a way for high GHG-emitting industries to avoid action," Scherr says. "There remains reluctance among some groups to fully utilize emission reductions and sequestration opportunities from the land-use sector."

This skepticism is likely reinforced by the fact that many agriculturalists—the people who would be doing the sequestering of carbon through better practices—profess a disbelief in human agency being behind global warming and only embrace it because selling carbon credits could become another income stream. But even before global warming became such an all-consuming issue, there was a wary distance at best between environmentalists and agriculturalists.

The conflict has its roots in America's very earliest concept of conservation, says Courtney White, an archaeologist who was a Sierra Club activist in the Southwest during the 1990s, when the conflict was at its most bitter. "There were rivers of bad blood between ranchers and environmentalists," White told me. "Booby traps were being set in the wilderness for loggers and people gathering wood. There was a bomb lobbed in the window of the Nevada Bureau of Land Management office. I thought, 'My God, what's going on?' I felt I had to do something."

The something that he did was to found the Quivira Coalition with a conservation-minded rancher who—to his surprise and joy—joined the Sierra Club. Quivira seeks the "radical middle" ground between agriculturalists and environmentalists and engages both in projects of common interest and education. A "do" tank, White says, as opposed to a think tank.

But he himself does a lot of thinking about the gulf between these two groups that have such a hard time talking to each other even though both love the land. In an essay for the Quivira Coalition journal, *Resilience,* he traces the evolutionary stages of the country's environmental movement, showing its historic divergence with people who worked the land for a living and the new and hopeful trend toward respect and collaboration.

Beginning in 1783, White writes, the policy of the US government was to encourage private citizens to settle on public lands and make productive use of them—to live the grand aspirations of what would later be called Manifest Destiny. That began to change in 1891, when president Ulysses S. Grant set aside Wyoming's Yellowstone as a national park. Nineteen years later, Congress created the National Forest Reserve system to protect valuable timberland from development and conserve it for future needs; president Theodore Roosevelt had doubled these reserves by 1907, only 4 years after he created the first National Wildlife Refuge at Pelican Island in eastern Florida. The National Park Service was founded in 1916 with about 35 parks and monuments; by the mid-1990s, it controlled more than 400 properties. After World War II, government extended its management of wild places when 175 million acres of public rangelands were put under the control of the Bureau of Land Management.

White describes all of this—including the creation of the Environmental Protection Agency in 1970 and environmental legislation in the 1970s—as the "federalist" wave of conservation. The underlying conceit was that government needed to play an active role in protecting the resources and even the wild beauty of America's iconic lands.

But that view of government as benevolent protector of public lands was becoming jaundiced in the 1960s and 1970s. Environmentalists bristled when they saw government agencies allow cattle grazing, logging, or mining on public lands. Ranchers and others who wanted working access to public lands bristled when government restricted and controlled their admittance, claiming that it was acting on behalf of urban-dwelling nature tourists and environmental kooks. The governmental agencies became less able to manage public lands effectively, partly because of dwindling budgets, but also because of growing dysfunction and resistance to change.

But the friction between environmentalists and agriculturalists had less to do with their view of government than their relationship with the land itself. Both loved the land, but for different reasons. Environmentalists thought the highest use of public lands and, indeed, the rest of the wild landscape, was to leave them alone and undeveloped, with the possible exception of recreational installations such as hiking trails and canoe concessions. And obviously, agriculturalists wanted to work the land. The industrialization of agriculture that had begun during the mid-20th century was also intensifying during this same period of time. Many small farms collapsed and sold their land to their bigger neighbors, who were following the exhortations from government and academia to get bigger still. Environmentalists could rightly point to a degradation of the environment as mega-farms took shape and their downriver impacts became obvious. But to many farmers and ranchers trying to hold on to their land and stay in business, the environmental ethos seemed like something they could ill afford.

Courtney White is familiar with the myopic views of

environmentalists, since he had been an activist himself. In his essay, he argues that their disregard for the livelihoods of rural people led to several problems. Their economic agenda of tourism and recreation had unforeseen negative effects, including congestion, pollution, and exurban sprawl. And they lost what Aldo Leopold, one of environmentalism's patron saints in the 1940s, called the feeling of "the soil between our toes," which White interprets as an intimate understanding of how the land actually works. Leopold had always insisted that people and their economic activities are part of the environment. "There is only one soil, one flora, one fauna, and one people, and hence only one conservation problem," White quotes from Leopold's 1935 lecture, "Land Pathology." "Economic and esthetic land uses can and must be integrated, usually on the same acre."

But in the 1990s, environmentalists were fiercely trying to pry the one from the other, and White found himself squarely in the middle. The contention was fierce in New Mexico, where environmentalists agitated for a ban on any form of logging on public lands, including traditional wood gathering for family use that Hispanic villagers had been conducting for centuries. "They really dropped the hammer on rural people," White told me. "The Hispanic community went kind of berserk, rightfully, because they saw this as racist. At one point, the villagers hung two prominent environmentalists in effigy."

The Sierra Club floated a referendum among its members calling for a ban of logging on public lands in 1996, and

it passed. A similar referendum banning ranching on public lands narrowly lost. White was a longtime environmentalist, but his background in anthropology also
informed the way he looked at the conflict. "I came out
of school thinking a lot about how people and culture and
history affect land," he said. "Any archaeologist can tell
you that past cultures have both benefitted and hurt their
environment. You can't be ahistorical and be a good conservationist. Unfortunately, many conservationists don't
look at it that way."

Indeed not. White opened the newspaper one day to
find himself being accused by a fellow environmentalist of
being *an archaeologist*—as if that were a badge of shame. "He
meant that I was interested in culture and history and people, whereas he was only interested in the environment,"
White said. "For me, that framed a lot of my discontent
with the movement."

These days, White thinks both the hard-core old-
school environmentalists and agriculturalists—the ones who
would never think of working with one another—are
increasingly irrelevant. "The rest of us have moved on," he
says. "We're trying to get local food systems going and trying to get carbon in soils and thinking about renewable
energy and so on. This larger group—I call it '21st-century
conservationists'—has a lot of things to think about."

These conservationists comprise both rural and
urban people, both agriculturalists and people whose only
connection to agriculture is what they choose to have for

dinner, and people working for government organizations and environmental groups. They have much in common, but one thing they often don't share is language. The verbiage surrounding climate change is highly charged and heavily politicized. Republicans used to talk about climate change in responsible ways, but were butted into denial by a well-funded campaign waged by the fossil-fuel industry. Al Gore stepped into the political vacuum but, say prominent pro-environment Republicans, did so in such a partisan way that it was easy for conservatives to scorn anything he championed. Given that most farmers and ranchers are deeply conservative, environmental activists maintain friendly ties with them by choosing their words carefully.

Feathers can fly when they don't, as Dave Nomsen, the vice president of government affairs for Pheasants Forever and Quail Forever found out.

Nomsen grew up in Iowa, which once had more than 3 million acres of prairie wetlands interspersed with various kinds of agricultural fields. Conventional corn and soybean agriculture has taken over the terrain and only 30,000 acres of those wetlands are left. One result is a loss of habitat for the game birds that Nomsen and other hunters prize. The disappearance of this once-congenial landscape made an environmentalist out of him, and he's been working since the late 1980s to get conservation into each consecutive federal farm bill.

In 2009, Nomsen wrote an article for Pheasants Forever's membership newsletter titled "Global Climate Change's

Inevitable Impact on Hunters and Wildlife." He detailed the changes taking place on the landscape that pierce a hunter's heart. With an increase in temperature, pheasant predators have expanded their range. Invasive species like fescue are edging out the native warm- and cool-season grasses that attract ground-nesting birds. "If you're a pheasant egg," Nomsen wrote, "even a one- or two-degree temperature rise or a change in grasses at the nest site can have a big impact on hatching rates. Couple changing microclimates at the nest site with more extreme weather patterns and a hen's ability to pull off a clutch becomes more complicated."

Nomsen tried to rally his hunter membership at the end of the article, pointing out that the habitat conservation they'd always pursued relieves the oversaturated atmosphere of carbon dioxide and sequesters it in soil and plants. "If global climate change means taking carbon out of the air through grass and tree plantings and wetland restorations, then Pheasants Forever has been in the global climate change business for over 26 years," he wrote.

Not what they wanted to hear! More members responded to this one article than to anything Nomsen had ever published, and 98 percent were negative. "We've got a lot of sportsmen and sportswomen who are very conservative in their values and thinking," he told me. "There is clearly the need for some additional education in this area."

Still, when it comes to work on the ground—on farms, on ranches, in wilderness areas and parks—the prescription for climate change mitigation is pretty much identical to the prescription for more productive working lands, clean water, better air quality, bountiful wildlife habitat, and so on. So many environmental organizations, from Defenders of Wildlife to the Environmental Defense Fund to Greenpeace, are now working with agriculturalists to build healthy, carbon-rich soils. They just don't talk about global warming, which has become like Voldemort in some circles—"He Who Must Not Be Named."

These collaborations work because these environmentalists don't march into rural areas and point to the people there and say, "Our waterways are getting polluted and *you* are the problem." Well, in some cases they do say something like that, but without the old hostility. For instance, Steve Richter from the Nature Conservancy (TNC) is working with a variety of partners to reduce the amount of phosphorus from fertilizers and sediment that is polluting the Wisconsin waterways from farmlands. The state government had discussed legislation to require farmers to buffer their croplands from the streams, but decided to try a carrot approach and forgo the stick.

Richter works with scientists to pinpoint how much phosphorus is leaving the fields worked by every farmer in the watershed, then approaches them to discuss it. He arrives with a handful of suggestions for ways to address the problem—using cover crops, going no-till, changing their fertilizer use, and so on—and offers to help them access federal conservation dollars to effect the change over a 2- to 3-year period.

"Some of them already wanted to try some of these practices, but this stuff has to work for them," Richter told me. "They want to build that soil and have it stay on their land. We're showing with this project that you can have these discussions and make change."

When I spoke to Richter, he was working with the 10 farmers whose fields drained the most phosphorus and sediment into the Pecatonica River. Nine had agreed to make changes on their land, and he had a meeting scheduled with the tenth.

A group of New England scientists formed the Nature Conservancy back in 1946—its original name was the Ecologists Union—to save biologically precious lands that were held privately. For nearly 40 years, the organization's primary focus was to purchase these lands, which they would either protect themselves or sell to the federal government for protection. By 2007, the Nature Conservancy was protecting 119 million acres of land and thousands of miles of rivers worldwide. They help protect approximately 15 million acres in the United States.

But this approach ignored the economic drivers of landscape degradation and failed to embrace the concept of working land. Bitterness festered after TNC's 1990 purchase of the historic Gray Ranch in New Mexico, considered one of the most significant ecological landscapes in America. The local population was angered by TNC's plan to sell this working ranch to the federal government. They protested the sale, and TNC listened. Instead of selling to the government, they sold the ranch to a local foundation established to maintain the property as a working ranch with a conservation agenda.

By now, the Nature Conservancy and many other environmental organizations operate in the same "radical center" as White's Quivira Coalition. Just as Steve Richter does in Wisconsin, environmentalists around the world are sitting down with farmers and ranchers to discuss their land management conundrums and look for solutions that dovetail with their environmental mission. Magnificent convergences are emerging piecemeal around the globe; hopefulness about our future seems reasonable in the light of these win-win ventures. This "new agrarianism," as Courtney White calls it, considers the vital thread running through both our physical and social landscapes. In this new vision, we can work with nature to heal the damage we've created. Farmers and ranchers can make a good and honorable living. And we can all eat food that's good for us, instead of the nutritionally vapid and often dangerous stuff that passes for food in our supermarkets.

In California, TNC has been buying orchards along the Sacramento and Cosumnes Rivers in the Central Valley. The rivers flood these orchards frequently, and water erosion shreds them. It's a win for the farmers eager to get capital out of these delicate areas and invest it elsewhere. It's a win for the environment overall, as TNC has restored some 5,000 acres of riparian forest along the Sacramento

River, buffering it from the orchardists' chemicals and allowing it to meander. The erosion that occurs naturally along the river's edge provides good habitat for the bank swallow, which burrows and nests into the freshly eroded banks every spring.

The organization also kept some of the orchards in operation, leasing them to local farmers and inviting them to experiment with using alternative and less chemically harsh ways of controlling pests. One of the harsh chemicals that TNC wanted to help farmers move away from was diazinon, a pesticide that attacks the nervous system—affected fish can't navigate properly. On the TNC orchards, plum growers figured out how to use pheromone mating disruption—insect sex hormones embedded in long, thin plastic tubes—to thwart pests. When they're attached to the branches of the trees, they flood the orchards with mating signals, which confuse the pests and make it hard for the males and females to zero in on each other. They fail to reproduce, and fewer pests swarm the orchard. TNC and the farmers conducted monitoring and held field days and eventually received the governor's award for integrated pest management. There's the other win: Those of us who eat these fruits and fish have one less hideous chemical to worry about.

TNC has also done a lot of work with nitrogen-fixing cover crops in California. Their test plots convinced many orchardists that this practice would not only provide nitrogen to their trees, but would also improve their soil and its ability to hold water, helping them cut back on irrigation costs. For a state that suffers chronic water shortages and that relies on mountain snowpack for 30 percent of its water—and that mountain snowpack faces an uncertain future in a warmer climate—anything that helps agriculturalists avoid runoff and hold water in the soil is a tremendous boon. "Cover crops really took off in the vineyards," said Dawit Zeleke, a TNC regional director and a mandarin orange grower himself. "In fact, a lot of vineyards were organic, but they didn't want to admit it because the organic sector of the wine business was so weak."

TNC and other environmental groups have also been working on problems related to rice agriculture, which occupies more than 500,000 acres in California and more than 2.5 million acres in Arkansas, Louisiana, Missouri, Mississippi, and Texas. Farmers used to dispose of rice field stubble by burning it—I remember looking over the Sacramento Valley from our house in Oroville, where the foothills of the Sierra Nevada began, and seeing a gray drapery of smoke hanging over the valley. I remember its faintly sweet odor. For people closer to the burn, though, the soot-laden air caused respiratory problems, and burning stubble was banned in 2000. Farmers now flood their fields after harvest to decompose the stubble in water.

When that new shimmer of water appeared in TNC's orchards, it turned out to be a beacon for migrating birds along the Pacific Flyway. Vast numbers of birds summer in Canada, Alaska, and the northern United States, and their ancestral routes to Central and South America have always wafted them into California's Central Valley, where many stay for the winter. But 95 percent of the valley's natural wetlands disappeared over the last century, leaving the birds precious little habitat as they seek their winter homes. The flooding of the rice fields presented them with a new freshwater option. As a result, their population has rebounded to healthy numbers, says Zeleke—including species like the white-faced ibis, which were nearly extinct.

And the birds are an unexpected boon for the rice farmers, too. The bacteria that they introduce to the fields through their droppings create a more diverse microbial population and more fertile, nitrogen-rich soil. The Nature Conservancy is now working with rice farmers to more purposefully manage their fields with the birds in mind: Together, they're experimenting with different water depths for different bird species, with draining their fields in stages so there's always some water for the birds, and with creating mudflats.

However, leaving the stubble on the ground to rot underwater created a new problem. This anaerobic decomposition creates methane, a greenhouse gas 34 times more powerful than carbon dioxide. This bothers the rice growers for a number of reasons, not the least of which being that it makes them a target for new and stringent state or federal regulations.

This is the kind of dilemma that attracts the Environmental Defense Fund (EDF). The organization has three different teams working on agricultural issues, and one headed by Robert Parkhurst studies science and fieldwork to figure out market solutions to climate change. Their overall goal is to cut 100 million metric tons of greenhouse gases from agriculture and forestry by 2020. That's about 40 percent of the annual greenhouse gas emissions from American automobiles.

Parkhurst's team began approaching rice farmers in 2007 to work toward a solution that benefits both agriculture and the environment. Working with a soil scientist, a research hydrologist, and a systems agronomist, EDF came up with a list of six practices that can reduce greenhouse gas emissions. Four of these practices reduce the amount of time the stubble as well as the seeds and growing rice are in contact with water during the warm growing season, when emissions are highest and the waterfowl have serendipitously departed. Some of these practices reduce the amount of water needed to grow the crop, which can create considerable savings for farmers in parts of the country where they run diesel engines to pump groundwater—one of their biggest costs. EDF wrapped these win-win practices into a protocol that was scheduled to become part of California's cap-and-trade program in the spring of 2014. "Producers can now farm rice and greenhouse gas offsets at the same time," Parkhurst told me.

Parkhurst's team is developing another protocol for an agricultural practice that not only reduces emissions, but also sucks carbon dioxide from the air and stores it in the soil as carbon. This practice employs one of the most ancient low-tech agricultural technologies on the planet, one that Rome's Pliny the Elder lauded in his writings: compost.

Elaine Ingham tells me that compost has a bad reputation these days. This shocks me, because I grew up in a household where compost was revered and nothing—nothing!—went into a garbage can that would have been useful for the compost pile. Or rather, compost piles—my parents had three bins in various stages of decomposition located under a deck off their living room. I liked dropping the cantaloupe rinds and carrot tops or whatever from that height. The rinds often hit the wooden sides of the bins and shattered. Carrot tops fluttered into lacy arrangements on top of the piles. Eggshells sometimes blew off course.

My parents had no idea why compost was so good for their garden; like most people, they probably assumed that that this broken down (mostly) plant material acted as a natural fertilizer. But that's not quite how it works. When you put a layer of compost over your soil, you're basically transporting a metropolis of tiny creatures to your yard. Any plants there will be serviced like one of the Tudor kings, with billions of microorganisms eager to exchange their gifts of phosphorus, nitrogen, and other nutrients for a sip of carbon.

But there is good compost and bad, so-called compost, Ingham says. The latter is often the stuff that waste control companies make from the land clippings and yard waste—and, here in Portland, kitchen scraps—that's picked up and carried away by loud trucks early in the morning. They make the bad so-called compost not to create a valuable soil amendment, but to shrink the amount of waste going into landfills. "They want to reduce a ten-foot pile into a two-foot pile," Ingham told me. "And sometimes they try to sell it to people as compost. They should stop calling it that. It's putrefied organic matter and it will kill your plants."

Good compost requires the aerobic decomposition of organic matter. Very basically, you assemble a pile of materials—some woody or dry plant materials (wood chips or dried leaves or even shredded newspaper) and some green plant materials (grass clippings) and a small amount of nitrogen-rich manure or legumes or nuts—and build it to a height of about 3 feet. Microorganisms live on the plant material and manure as well as in the soil, and they begin to eat the simple carbons in these piles and poop out slightly more complex carbon chains and reproduce. The heat of their reproduction causes the pile to warm—and yes, when you stick your hand in a compost pile or even a heap of grass clippings and feel the heat, that's the hubba-hubba of microbial sex.

A good compost pile heats up so much that it kills pathogens and weed seeds. But Ingham says that if the interior of the pile reaches 160 degrees, it needs to be turned. All that microbial activity requires oxygen, but if the heat rises past 160 it means that the microbes are reproducing so quickly that they will use oxygen faster than it can penetrate the pile. If that happens, the aerobic microbes go to sleep and the anaerobic microbes take over. Then, instead of a sweet-smelling compost hued, Ingham says, like a chocolate bar with a 70 percent cocoa content, you'll have a foul-smelling black pile of so-called compost filled with ammonia and other nasty anaerobic by-products.

Good compost can work wonders. Berkeley soil scientist Whendee Silver and her students have been conducting research on the power of compost at a 539-acre cattle ranch on the California coast near Nicasio. As part of a collaboration called the Marin Carbon Project, they spread a centimeter of compost on the soil and monitored the site for 3 years to observe changes above and below the soil. The changes above the soil were apparent to even the nonscientist: The grasses exploded in a green flurry of growth, for a 50 percent increase in forage. Underground, Silver discovered, the amount of carbon in the soil was also increasing dramatically. Overall, the carbon sequestered in the plants and ground increased by 25 to 70 percent at six sites, *not* including the carbon in the compost itself.

Based on her research, Silver estimates that just half of California's 63 million acres of rangeland could absorb 42 million tons of CO_2—nearly 40 percent of what California's power plants emit in a year—if each acre sequestered an additional 1.5 tons of carbon in its soil. "Given what we've seen in our experiments," Silver told me, "one and a half tons is doable."

Silver and her students made their compost from city yard waste mixed with agricultural waste, including manure. While it sounds impractical to truck hedge trimmings and spent dandelions all the way out to a ranch so they can join cow pies in a compost pile, the Environmental Defense Fund, which is a partner in the Marin Carbon Project, believes an approach like this would work well in certain areas. Large municipalities generate a lot of food and green waste, and many are surrounded by rangeland. When dumped in landfills, these organic wastes emit methane. But if this organic waste is diverted from the landfill, then composted, then applied to nearby rangelands, the overall impact on greenhouse gases could be significant. And not just in California, where Parkhurst and his team have turned this work into another protocol, first for Belinda Morris's American Carbon Registry for the voluntary market and thereafter for the California Cap-and-Trade Program. Fifty percent of California's land is rangeland, but a third of the United States is also rangeland. What if compost could work in other areas, too—New Mexico or Colorado or Arizona or Montana? If that were the case, spreading compost on rangelands could be *very* big.

"It's not a silver bullet," Parkhurst told me. "It's not going to work everywhere for everyone, but it will be a solution to the problem. The sources of climate change are many, and so the solutions have to be just as many."

What makes all this work possible, Parkhurst hastened to add, is the new wave of effort by scientists who are finally paying attention to the life in the soil. "The interaction between microbes and plants and water and carbon is something we're just now really beginning to understand better," he said. "That's what allows us to calculate the amount of emissions averted or absorbed."

Where the Environmental Defense Fund sees an opportunity to create models and carbon-offset protocols, the World Wildlife Fund (WWF) sees an opportunity to influence supply chains. WWF has a team that focuses on key commodities from around the world that affect conservation. One of these commodities is beef. The organization's main focus in the United States is in the northern Great Plains, one of the few intact grasslands in the world. The Great Plains blanket the land with grass from the Missouri River to the Rocky Mountains, bounded by Saskatchewan on the north and the Nebraska Sand Hills on the south.

"These grasslands are intact because of the ranching community," said the WWF's Jeff Nelson. "We want to see these ranchers have sustainable enterprises. If they struggle and leave the land, what replaces them might not be nearly as good for conservation."

Condominiums and cornfields—both are fetching high prices these days, tempting ranchers to either sell or plow up the carbon that's been accumulating in their soils for centuries. So the WWF is trying to focus on ways of compensating ranchers for the carbon they're already protecting in the soil, as well as the wildlife they can shelter through improved practices. WWF is optimistic they can do this, because they have the most powerful ally in America on their side.

That would be us, the consumers.

In the early 2000s, a groundswell of voices began to extol the virtues of cattle raised on grass—the food they evolved to eat—as opposed to corn, the food they're obliged to eat in modern feedlots and in many dairies. Corporate agriculture switched cows from greens to grain because they bulk up faster: One study has shown that cattle eating grasses and other forage gain only three quarters the weight of cattle raised on grains. But that weight gain, which translates into heftier profits for the likes of ConAgra and National Farms, comes at a cost to the animals' health. Corn stresses the bovine digestive system and can lead to bloat and other disorders, all the way up to liver failure.

This steady diet of corn along with the cramped quarters in feedlots and confined dairies tends to keep cattle in chronic poor health, so the operators routinely mix antibiotics into their feed. Antibiotic-resistant bacteria can develop as a result and then can be passed on to humans through beef consumption and leave us vulnerable when we're sick and need the intervention of antibiotics.

Journalist Jo Robinson, author of *Eating on the Wild Side* and *Pasture Perfect,* was one of the earliest advocates for putting cows out to pasture. Her Web site Eatwild.com collects the studies underlining the differences between the meat from grass-fed and corn-fed cattle. Among these studies:

Researchers from the Translational Genomics Research Institute (TGen) tested meat from US grocery stores and found that nearly half were infected with *Staphylococcus aureus* ("staph") bacteria, which are closely linked to several human illnesses. The staph found in the meat was resistant to three classes of antibiotics. "Densely-stocked industrial farms, where food animals are steadily fed low doses of antibiotics . . . are ideal breeding grounds for drug-resistant bacteria that move from animals to humans," TGen's Web site says.

On the other hand, meat and milk from cattle raised on pasture seem to confer a number of health benefits. Robinson's survey of the scientific literature points to four major pluses.

First, pasture-raised meat has less fat and calories overall, since the animals are able to move around. Potentially more important, the quality of the fat from grazed cattle is quite different in that it has two to four times as many omega-3 fatty acids as grain-fed beef. These fats, which come from the grass itself (just as the omega-3s in wild salmon come from the green algae eaten by their prey lower on the food chain), "play a vital role in every cell and system in your body," Robinson writes, and summon an armada of health benefits: lower risks of various cardiac problems and cancer, as well as lower rates of depression, schizophrenia, attention deficit disorder, and Alzheimer' disease.

The meat from pasture-raised cattle is also higher in vitamin E, which may have antiaging properties and lower risks of heart disease and cancer. Meat and dairy products are also the best known source of another good fat called conjugated linoleic acid (CLA). "CLA may be one of our most potent defenses against cancer," Robinson writes. "In a Finnish study, women who had the highest levels of CLA in their diet had a 60 percent lower risk of breast cancer that those with the lowest levels. Switching from grain-fed to grass-fed meat and dairy products places women in this lowest risk category."

When picky consumers and healthy-eating gurus started the clamor for grass-fed meat, many cattle producers were aghast that anyone expected them to return to the practices of their fathers and grandfathers. "Jo Robinson was a speaker at one of our early conferences, and her talk was quite controversial," Quivira's Courtney White told me. "One rancher said, 'You can't say that about feedlots!' Fast-forward, and 5 years later he was producing grass-fed beef himself. A lot of conservative, hard-nosed ranchers who said they'd never do it are now doing it."

So the mostly urban–based food movement is already helping to green the countryside, as the higher prices fetched by grass-fed meat and dairy convince agriculturalists to put their animals back out on pasture. The World Wildlife Fund hopes to attach even greater rewards to this trend. When I talked to Jeff Nelson, the organization had convened a first meeting in Montana with some of the state's largest ranchers, retailers like McDonald's and Wal-Mart, which are feeling the heat from consumers, and huge commodity groups like the National Cattlemen's Beef

Association. In effect, this meeting and others that followed collapsed the retail chain from the pasture through the packers and processors all the way to the checkout register. The goal is to create a certification program for beef that meets certain standards all along the way.

Financial rewards are often the hook that draws agriculturalists closer to the environmental camp, but changed hearts tend to sprout unexpected wings.

The conference I attended in Davis was convened by the California Rangeland Conservation Coalition, a group formed in 2005 by unlikely allies. On the one side, the Defenders of Wildlife (DOW) were trying to protect vernal pools, an ephemeral wetland habitat found largely in grasslands that typically forms from spring runoff and snowmelt and dries up in the summer. While they're wet, vernal pools are an important ecosystem for amphibians and insects that can't survive in fishy waters, but they were being bulldozed for housing developments all over California. DOW wanted the state's fish and wildlife service to protect these fragile areas, and someone at the agency suggested they seek allies in the ranching community. Ranchers were also worried about developers turning grasslands into cul-de-sacs, as it made land prohibitively expensive and threatened their entire way of life.

Don't cattle damage fragile habitats? That was certainly the common assumption among environmentalists, but research published by Nature Conservancy scientist Jaymee Marty in 2005 proved it wrong. Over the course of

3 years, Marty observed that grazing actually helped vernal pools thrive by releasing the chokehold of exotic invasive plants. When she compared grasslands with no grazing to ones with continual grazing, she found that the ungrazed vernal pools had 88 percent greater cover from annual invasive plants and 47 percent lower cover from native species. Ungrazed, the species richness of native plants declined by 25 percent and the diversity of aquatic insects was 28 percent lower. Overall, the period of time these fleeting wetlands stayed wet dropped by 50 to 80 percent without grazing, making it difficult for some species that lived there to complete their life cycles. "We wouldn't have gone ahead [with this partnership] without the science showing that managed grazing would actually benefit the vegetation communities and wildlife we wanted to protect," said Kim Delfino, DOW's program director in California.

So these two groups began a careful tiptoe toward each other. The first formal gathering was a barbecue at a Sunol, California, ranch that included ranchers, DOW, the Environmental Defense Fund, the Endangered Species Coalition, the California Cattlemen's Association, the US Forest Service, and others. "For the first 30 minutes, it looked like an eighth-grade dance with boys on one side and girls on the other, staring at each other," rancher Tim Koopmann told the University of California at Davis magazine. "Except in our case, the environmentalists stood against one wall and the cattlemen against another. But then we started talking and knew we could make it work."

Now, they refer to themselves as the Boots and Birkenstocks Bunch. They have a conference every year that looks at the environmental benefits of grazing as well as the grazing practices that most enhance the environment. Together they work on delivering conservation dollars to good ranchers and getting state and federal regulations in place that support both conservation and the cattle industry. The partnership is so strong that ranchers and environmentalists can now broach even issues that would have had them at loggerheads before. "Wolves are coming from Oregon into California," Delfino told me. "We've been able to talk to the ranching community about how to deal with that so much more easily than our colleagues in Wyoming or Montana or Idaho."

When Kansas rancher Bill Sproul told me that he and DOW's Jessica Musengezi would have sat at the same table 10 years ago only if one was suing the other, he wasn't speaking from personal experience. Sproul is a first-generation rancher who perhaps didn't have any ancestral suspicion of environmentalists, and he was never opposed to the idea of conservation. He just wanted to make sure he was paid for it.

"I used to think on the commodity level," Sproul told me. "I owned things and I used things and I made money. If you wanted me to do conservation, fine, but how much could you pay me to do it? What can I make out of it?"

But the late 1990s took him to a triangular chunk of land spanning southeast Arizona and southwest New Mexico, where a trial by fire had turned longtime headbutters into partners. Earlier that decade, the Forest Service had rushed to put out a 500-acre wildfire on private land over the objection of the landowner. The Forest Service had a policy of extinguishing all fires, but ranchers and some scientists felt that fire was a natural feature of their ecosystem and that suppressing it had allowed brush to take over too much of their grasslands. Leaders in the ranching community had already been meeting to discuss other challenges, including the controversy over ranching on public lands that Courtney White found so distressing. This confrontation with the Forest Service made them realize that a policy of digging in their heels alone was useless. This was the genesis of the Malpai (named for the volcanic rock in the area) Borderlands Group, a pioneering coalition of ranchers, scientists, conservationists, government agencies, and concerned citizens. Rancher Bill McDonald, a member of the group, coined the term "radical center" which now informs and inspires similar groups all over the country.

Since then, Sproul has moved toward what he calls community conservation. "Everything is part of the community," he said. "I'm part of it, you're part of it, the cattle are part of it. The air, the land, the ants, the lizards. And the whole idea is to practice conservation on a landscape scale. Wildlife and clean air don't know where ranch boundaries are. They don't know where Kansas ends and Nebraska starts."

Sproul feels that this perspective ultimately benefits him, because his cattle are healthier when the landscape thrives. But sometimes he takes a financial hit to benefit this community. Like many ranchers in his area, he burns his grasslands every year to get a new flush of grass for his yearling steers, which he only keeps for 90 days before they're sent to a feedlot. By doing this, his cattle gain between a quarter and a third more pounds per day, and that adds up to a considerably larger sum in his bank account at the end of the 90 days.

On the other hand, burning *all* the grass means that birds that nest in the previous year's old grass lose habitat. The number of grassland birds is declining, and he suspects there's some correlation. So he's changed his annual burn to leave a mosaic of dead grass for the birds.

"I take a bit of a hit in the checkbook, but I'm okay with it," Sproul told me. "Healthy isn't just what's in my checkbook. Healthy is also that scissor-tailed flycatcher sitting out there on that fence, looking at me."

CHAPTER 8

HEROES OF THE UNDERGROUND

Inoticed it over and over again as I worked on this book, whether I was talking to farmers or ranchers or scientists— a quality so often missing in the ongoing dialogue about our world that I at first mistook it for happiness, the sheer pleasure one takes in work that one loves.

But it was more than that. It was optimism.

I offered up this observation about the soil-health visionaries I'd met to Peter Donovan, the piano tuner/ musician, soil autodidact, and conductor of the Soil Carbon Challenge.

"They're connected with the most powerful geologic force, which is life," he said. "Most of what we take to be the physical environment is, in fact, the creation of living organisms over time. That's a different paradigm than the idea that life is a fragile passenger on a dead planet."

When he said this, we were sitting in his bus parked on a dirt road that bisected the farm, herbs growing on his hood, pasture burgeoning in the fields, the busyness of the sheep and sheepdog a distant noise, the cheerful agrarian commerce of the Corvallis farmers' market yet to come and much anticipated. I liked this idea of the robustness and ongoing creativity of life, and it was easy to feel surrounded by it there.

But what about back in the city?

We often have the idea that cities are dead zones—concrete jungles—where the only life is that of us humans, moving through our various intersecting orbits, along with the rats, pigeons, and cockroaches that flourish in our presence. But cities are also teeming with life. Microbial ecologist Peter Groffman of the Cary Institute of Ecosystem Studies in Millbrook, New York, says that only 20 percent of the urban landscape is truly impervious. The other 80 percent is natural or seminatural. "There are really a lot of plants in urban areas, a lot of animals, and a lot of ecological function," explained Groffman, who since 1998 has been studying soil and water in Baltimore, one of 26 long-term ecological research sites funded by the National Science Foundation. "There is a lot of soil functioning in urban areas"—the buildup of carbon, impeding nitrogen and phosphorus from washing away—"and, to me, that's a really positive message."

I didn't really need Groffman to tell me this, since I've never thought of cities as dead zones. For the last 20 or so years, I've walked my dogs along the same morning route, first in Cleveland and now in Portland. There are at least two good things about a fixed itinerary. Since I never have to think about where I'm going, my mind is free to wander. And it wanders and wonders, quite frequently, over the daily changes in the yards, parks, tree lawns, and gutters around me.

Only recently, I've seen how orange poppies burst forth in the spring, fade, and are displaced by pink-eyed phlox. Trees bloom and fruit, and sometimes all that progeny is too much for the tree to bear: A branch breaks, the fruit spills, and people squash it into boozy brown pulps on the sidewalk. After what seems like just a month, weedy upstarts in pavement cracks grow into sunflowers the size of saplings. A vacant lot is a swirl of purple sweet peas, blue chicory, and wild fennel that makes my dogs smell wonderful. Then it's mowed and features cans and discarded plastic forks instead of flowers.

I was curious to see how all that I had learned about the amazing life in the soil fared in these urban settings. In Cleveland, my dogs and I used to walk past huge houses surrounded by immense landscaped yards. Gardening season always brought funky air, as trucks arrived to spray lawn chemicals. When the rains came, muddy water often ran off and left a slick chocolate scrim on the sidewalks. I'd survey the offending property for evidence of exposed soil, such as unmulched beds, trees surrounded by wide bare circles, and land rototilled and scraped clean for vegetable plots. I imagined the deprived soil biology in these naked spots shrieking, "Feed me!" like the voracious plant in *Little Shop of Horrors*. For the living things in the soil, these barren patches are food deserts.

Portland has built swales along many streets to catch

runoff. Still, I see lots of the same blights on the soil in people's gardens. Plus, there is an unholy fondness for gravel in Portland: whole tree lawns, even whole front yards covered with stony matter. Perhaps because plants are too much bother? Or because people think that gravel provides better erosion control than roots?

Groffman and other scientists are taking a serious look at ecological function in cities. He's excited about the trend of urban agriculture—farms are springing up in the wake of foreclosures in places like Cleveland—but also about all the possibilities offered by that previously unrecognized 80 percent. "People manage these areas," he said. "We live and work there. If we learn how they function, then we can improve or alter that capacity to reach certain goals, whether it's carbon storage to regulate the climate or absorbing water. It's a great opportunity to use our understanding to achieve specific goals."

The challenge seems to lie in what we require of that 80 percent: We want the urban flower beds to pack a lot of pretty into well-defined slivers, and we want the turf to be durable enough for dogs, children, and adults wielding everything from strollers to soccer balls. Historically, we've been persuaded that synthetic chemicals and lots of gas-guzzling tools are needed to attain these goals.

But just as the agricultural world boasts pioneers and outliers, so does the urban world. One such is Eric T. Fleisher, director of horticulture at New York's Battery Park City Parks Conservancy. Fleisher manages the park's

36 acres organically, without either synthetic fertilizers or pesticides. His job is a demanding one, and not just because his efforts have to please the public. His work also has to *withstand* the public. By the park's last official count, up to 15 million people visit in a year, but Fleisher says they figure the number of visitors might now be as high as 25 million. That's a lot of impact by sneaker and sandal, especially in the playgrounds. Even the vibrations of the subway increase soil compaction. But Fleisher and his staff make the place lush and gorgeous without synthetic chemicals.

They do this by respecting the ancient synergy between plants and soil organisms. They test the park's soil every year and much more often in problem spots, examining the composition of the soil's biology. If any crucial players are missing, Fleisher reintroduces them with different recipes for compost and compost tea made from waste materials generated within the park. He's emphatic that he's not making the compost to reduce that waste and the cost of hiring someone to haul it away. "It's one thing to compost this material just to get rid of it," he told me. "It's another to be producing something that feeds the nutrient cycling in the soil."

Even if soil is rich with nutrient-toting bacteria and fungi, plants can't access the nutrients unless predator protozoa and nematodes feed on the bacteria and fungi and excrete their nutrient treasures in plant-palatable forms. Fleisher and his crew have found that they can cure their compost in ways that produce greater populations of these predators—warmer and dryer piles produce hordes of nematodes and cooler and wetter piles cause the protozoa population to boom. Before application, they put their composts and teas under a microscope to determine who's there. With this kind of precision, they can fix chemical deficiencies in the soil.

Fleisher's organic management of the park has led to healthier plants and healthier soil with greater water-holding capacity. That means not only that there's less need for irrigation, but also that torrential downpours don't turn the playgrounds into ponds. They've created the kind of healthy soil that soaks up the water quickly and holds it for a long time.

Change is frustratingly slow—really, why isn't every manager of large urban parks and gardens doing this?—but Fleisher's influence is spreading. In 2008, he spent a year as a Loeb Fellow at Harvard's Graduate School of Design and, while there, helped Harvard's landscaping staff apply organic principles to a test plot in Harvard Yard. The test was so successful that Harvard now manages 85 acres organically, creating compost and teas out of waste material from the site as well as cafeteria and dining hall leftovers. In a 2009 article for the *New York Times,* Anne Raver reported that the soil in Harvard Yard—where some 8,000 people pass daily—was previously so compacted that the trees were dying. Now, the soil is so healthy and well structured that the trees' roots find plenty of room to expand for water, oxygen, and nutrients. The landscaping staff also saved a

40-year-old orchard there from leaf spot and apple scab by applying local compost tea.

Fleisher helped Boston's Rose F. Kennedy Greenway launch organically and is now helping Princeton University transition to organic, and it's possible that managers of city landscapes across the country may have new motivation to seek the organic approach. Peter Groffman told me that many Eastern cities are developing tree canopy goals (New York City's is to plant 1 million more trees), since they cool cities in the summer, warm them in the winter, and filter pollutants from the air. And they're beautiful!

"There are all these ecosystem services that come from trees, and cities want to plant more of them," he said. "To do that, they fundamentally need to understand what's going on in urban soils and how to amend them to get trees to grow. They don't want to overfertilize soil, or they'll end up with more water and air pollution."

On my morning walk today, I passed a young tree planted on a sun–bleached tree lawn (it hadn't rained in weeks, and many Portlanders just let their lawns go brown in the summer). An oval of dirt had been incised into the grass around its thin trunk, and the homeowners had taken some care to remove every blade of grass within. The soil there was as hard and flat as a cutting-board, and my first thought was that the tree didn't stand much of a chance. But Groffman's research in Baltimore has led him to some startling conclusions. "We assumed lawns were going to be biological deserts," he told me. "They're not. We're taking core samples down to a meter and we're finding a lot of natural-looking soil profiles in these lawns, not just compacted fill. A lot of roots, a lot of biology, a lot of carbon, and a better environmental performance than we expected."

We might be able to sniff lawn chemicals from a block away, but that doesn't necessarily mean lots of people are using them. Groffman found that homeowners aren't managing their turf as aggressively as assumed. His surveys show that only about half of homeowners apply fertilizer to their lawns, and most don't fertilize at a high rate.

"Some people have a real environmental ethic and don't want to be adding pesticides and fertilizers and weed killers to the environment," he told me. "There's also, for lack of a better term, just laziness. People do the minimum to keep their lawns looking nice enough so their appearance doesn't bother their neighbors."

With fewer chemicals to turbocharge the performance of some plants and kill others, lawns become more diverse environments, both above and below the soil line. And just as in nature and on well-managed farms and ranches, this diversity supports the ancient and noble partnership of plants working with the biology in the soil. That partnership snatches carbon dioxide from the atmosphere, breaks it down into carbon, and puts the latter to good use down under.

About 80 percent of American homeowners have lawns. How each of us manages our own individual turf may seem insignificant to the greater picture, given all the factors that influence our world, but our turfs and turfettes add up. Lawns are the largest irrigated crop in America, taking up three times as much space as corn, the next-biggest irrigated crop. What we do with our urban green matters, whether it's in our yards or our parks or even our highway median strips.

Are the lawns managed by the 50 percent of home owners who use no synthetic chemicals less pleasing to the eye? It depends upon the eye. When Bob Streitmatter took over management of Luthy Botanical Garden in Peoria, Illinois, he quietly took the place organic. Clover began appearing in the turf grass, and Luthy's visitors were perspicacious and picky enough to comment on it. But he quickly won them over, explaining that clover is a legume that makes nitrogen available to the turf and allows the staff to do away with fertilizer. Allowing clover to grow among the grass not only saves the public from synthetic chemicals, but also saves money that could be spent on other programs.

"When you start making the change, it can be a little more costly," Streitmatter told me. "But you're setting up a soil ecosystem for the long haul. Down the road, your costs drop dramatically."

Luthy's transition was so successful that Streitmatter now runs a class teaching homeowners how to transition their own yards to soil health. He points out that a lot of these seemingly new ideas—like lawn care without chemicals and even rain barrels—were fixtures in the past. Clover actually used to be planted in partnership with turf grasses before monoculture lawns became vogue, and the sight of it reassured homeowners that their lawns were healthy. Then Americans were swayed by the marketing of the landscaping chemical companies and started thinking that their lawns should look like the course at the Augusta National Golf Club.

"I tell people that they aren't going to be hosting the Masters," Streitmatter said. "There's no need to put that much water, chemicals, and labor into their lawn. It's just not necessary."

Streitmatter also guides homeowners to make other choices in their yards to create soil health, and these echo the steps progressive farmers and ranchers are taking to build soil health on their land. He recommends flower beds designed for maximum biodiversity and density, featuring shrubs, perennials, and bulbs for both spring and fall bloom, with some annuals and biennials wandering through. He suggests tucking some leguminous plants in there to add nitrogen. All these different landscaping plants will feed slightly different recipes of carbon exudates to the microorganisms below—different cakes and cookies, as Elaine Ingham calls them—and ensure that diverse biology thrives underground. Planting them densely not only increases the amount of exudates, but also protects the soil from erosion and carbon loss. And as much as possible, Streitmatter tries to combine plants as they occur in native plant communities. His thinking: It's possible that these plants don't grow near one another just by chance, but because they benefit one another in some yet-unknown way, just as good neighbors do everywhere.

Instead of applying herbicides, Streitmatter suggests hand pulling weeds and heavily mulching large weedy areas with "lasagna" compost—alternating a layer of brown or dry stuff, like dead leaves and newspaper, with a layer of green stuff, like mown grass and plant trimmings—an easy way to discourage weeds, maintain soil moisture, and increase biological activity. He warns against cutting grass lower than 2½ inches—Battery Park goes no lower than 3 inches—to protect the turf plants and discourage weeds. And there's no need to plant our seeds or seedlings in spindly little rows separated by broad strips of bare soil when designing a vegetable garden! Instead, Streitmatter encourages gardeners to sow vegetable seeds in masses—scatter them by the square foot, rather than the row—so that the growing plants cover the bare soil completely. I've done this in my new little raised bed, just snipping away the tiny growing arugula and kale when I want to harvest. I leave the roots there for my underground herd of microorganisms to polish off. Cucumbers are rooted around my tomato plants and seem to be growing happily in their shade. I've never planted my vegetables so densely, but it seems to be working: I'm having a good harvest from this box of nutrient jungle.

I'm trying to be a hero of the underground, in my own small way. We can all be heroes of the underground by taking care of the soil around us, patronizing the agriculturalists who take care of theirs, and monitoring the political climate that affects soil health around the world. We have to pay attention to the farm bill! It's the single largest manifestation of our nation's food policy, and it has a huge impact on how our food is grown and how agriculture affects the greater environment, and on who benefits from this enterprise. Will it be consumers and average farmers, or will it be agribusiness? It helps to heed the work of organizations like the Union of Concerned Scientists, Food and Water Watch, the Environmental Working Group, and the National Sustainable Agriculture Coalition, which can help us understand the issues behind the farm bill and other food policy measures. It's too hard to understand this on our own. And with that knowledge, we should take whatever action we can—calling our elected representatives, voting, protesting, whatever.

Can we work with nature—and her so-called low technology, although humans have nothing that compares to its complexity and sophistication—to pull the carbon dioxide that we've overloaded into the atmosphere and make good use of it in the ground? Can we reverse global warming in time to leave our children a good earth?

So many smart people are working on this that I'm hopeful. Just as I was finishing this last chapter, Peter Donovan e-mailed me a presentation made to the Sandia National Laboratories in New Mexico by the New Mexico State University (NMSU) Institute for Sustainable Agricultural Research. As of this writing, it may represent the most exciting science on this subject.

The research began when NMSU molecular biologist

David C. Johnson conducted an experiment for the USDA to develop a process for making lower-salinity compost from cow manure (apparently, cow poop is salty—maybe from all the salt licks?). Johnson and his wife, Hui-Chun Su Johnson, did finally develop a compost with lower salinity, although he's not exactly sure why it's lower. His working assumption is that they made the compost using a special "no-turn" method—most composts are turned when the pile gets too hot to prevent it from going anaerobic—and that left the fungi populations undisturbed and able to flourish, while still admitting enough oxygen for the beneficial bacteria. He thinks it's possible that the fungi tie up the salts and protect the plants.

Johnson then tested the performance of this compost plus eight others on chile plants in a greenhouse. The work really became interesting when he noticed that the chile plants grew with twice the vigor in his compost. When he teased out the various factors that affected growth among these greenhouse plants, he found that none of the nutrients that are typically considered essential to plant growth—in particular, nitrogen, phosphorus, and potassium, which are the primary ingredients in conventional synthetic fertilizers and which were naturally present in all the composts—was the most important factor. Even soil organic matter—the non-rock part of the soil that includes bulky plant and animal material all the way down to highly concentrated humin—was not the primary factor. Instead, Johnson's compost had a balanced population of healthy fungi and bacteria, compared with the predominantly bacterial population in the other composts, and it was this balance that made the chiles thrive. Typical soil tests, which look only at the chemistry in the soil and not the biology, will miss this.

Intrigued, Johnson joined with NMSU entomologist John Ellington and engineer Wes Eaton to take this work into outside test plots, where they built up the soil biology—this time by using cover crops. They didn't clear the land for replanting; instead, new crops were planted within the debris of the last season's cover crops. At the end of 2 years, the soil organic matter leapt by 67 percent and the soil's water-holding capacity jumped by more than 30 percent—a big deal in this arid climate. The best of these plots produced four times as much green stuff as the world's most productive ecosystems.

"We showed that you can grow more crops faster, better, and with less water on soils where we've improved the population of microbes, both fungi and bacteria," Johnson told me. "The carbon sequestration is the icing on the cake."

The NMSU research suggests that the soil really can save us—and faster than anyone expected. In these plots made highly fertile by the interaction between plants and the soil biology, the plants were generously shuttling 72 percent of the carbon they pulled from the air into the soil. Even more amazing, far more of this carbon was staying fixed in the soil compared to land where the soil biology is weak and maltreated. While the researchers expected the amount of CO_2 wafting off the soil—the exhaled breath of the soil biology—to increase as the microorganisms themselves increased, the rate of respiration actually dropped. Meaning, soil carbon storage was accelerating in a nonlinear fashion. Two plus two was adding up to 15 or 20.

To explain this, Johnson suggests thinking of the energy needed to build soil carbon as being like the energy needed to get an airplane off the ground. There is initially a heavy investment of energy to get the plane in the air. But once it's up, the resistance is lower and the thing *flies.* In the same way, once that population of soil microorganisms is established, there are greater efficiencies in both growing a crop and growing soil carbon.

Weirdly, we've all been schooled in the notion that plants are takers, removing nutrients from the soil and leaving it poorer. But when plants are allowed to work with their partners in the soil, they're givers. They feed carbon exudates to the community of bacteria and fungi to keep them thrumming with life and pulling mineral nutrients from the bedrock as well as from particles of sand, silt, and clay because they know—if that word can be applied to organisms without brains—that they will profit from the gift. When the predator soil organisms eat the bacteria and fungi, all those nutrients are released near the plant. There's always enough, unless humans or some other force messes up the system.

One of the conclusions Johnson and the other NMSU researchers reached might make the toes curl on climate activists around the world. In their report to Sandia—delivered in the context of a project to see if the Mars *Curiosity* rover's carbon-measuring equipment could be used for quick and accurate analyses of soil carbon on earth—they stated, "The rates of biomass production we are currently observing in this system have the capability to capture enough CO_2 (50 tons CO_2/acre) to offset all anthropogenic CO_2 emissions on less than 11 percent of world cropland. Over twice this amount of land is fallow at any time worldwide." Meaning that if only 11 percent of the world's cropland—land that is typically not in use—improved its community of soil microorganisms as much as Johnson and his colleagues did in their test plots, the amount of carbon sequestered in the soil would offset all our current emissions of carbon dioxide.

It's a rather staggering assertion.

"Aren't you afraid to say this?" I asked Johnson. "Aren't you afraid saying that will let the oil and gas companies off the hook? As well as the people burning down forests and all the rest of us with a big fat carbon footprint? Aren't you afraid?"

I thought I could feel a wary shrug over the phone. "I don't see anything on the horizon that touches the effectiveness of this approach, plus it has so many other benefits," he said. "We're not going to reduce our carbon dioxide emissions anytime soon, because we depend too much on oil and gas, and the rest of the world wants our lifestyle. The whole idea is to get something that works right now,

234 // THE SOIL WILL SAVE US

the world over, to make a significant impact on reducing atmospheric carbon dioxide."

He added that we aren't yet scientifically certain that removing CO_2 from our atmosphere will reverse the changes in our climate—we've never done it, and even the best models can be wrong. Still, he says, it would be foolish not to try to do it by building healthy soil, since this also improves plant productivity, allows us to use less water, reduces our use of dwindling natural resources like oil and phosphorus, and lowers the impact of agriculture on the environment. Back to the question posed by so many apologists for industrial agriculture, How do we feed 9 billion people? Answer: Let's begin by feeding our microbes.

It appears to be in our power to reduce our legacy load of carbon dioxide, which has a physical presence as well as psychic weight. It's a question of national will and priorities. We can only do it by working with plants and soil microorganisms, which have been carrying on the most wondrous dance since the early morning of time. We can't keep being the oaf that breaks into the dance, bumping one partner or the other out of the way, thinking we can improve upon their step and sway. We suffer for this clumsiness. We need to stand back, pay close attention to the ways in which these partners need our help, and offer it with the greatest of respect.

ACKNOWLEDGMENTS

A nonspecialist who undertakes a book like this winds up saying "please" and "thank you" over and over. They can't be said enough! I am everlastingly grateful to the people I interviewed and, in some cases, observed, for their kindness and patience, and their enthusiasm for the subject. And really, I couldn't believe my luck that someone was paying me to talk to them—they are brilliant and brave pioneers, most of them toiling against great odds to bring a new way of understanding the physical world into the light. Some did yeoman's work in helping me translate complex information into narrative, not only explaining their work over the course of many interviews but also critiquing and correcting parts of the manuscript and answering dozens of follow-up questions by email and text, even at weird hours. In the pantheon of these super sources are Carl Bauer, Eliav Bitan, Gabe Brown, Jody Butterfield, Adam Chambers, Abe Collins, Peter Donovan, Eric T. Fleisher, Jim Fleming, Jay Fuhrer, Rick Haney, Elaine Ingham, David C. Johnson, Christine Jones, Fred Kirschenmann, Rattan Lal, Craig Leggit, Jonathan Lundgren, Belinda Morris, Kristine Nichols, William Ruddiman, Ricardo Salvador, Alan Savory, Bob Streitmatter, Courtney White, Bob Willis, and Dawit Zeleke. I am also grateful to Pelayo Alvarez, Steve Apfelbaum, Ricardo Bayon, Matthew Benz, Parker Bosley,

Valerie Calegari, Mike Callicrate, Wayne Carbone, Jerome Chateau, Cynthia Cory, Craig Cox, Dorn Cox, John Crawford, Kim Delfino, Randal Dell, Cornelia Flora, Brian J. Ford, Stuart Grandy, LaVon Griffieon, Peter Groffman, Alan Guebert, Ian and Di Haggerty, Sharon Hall, Wayne Hanselka, Michael Hansen, Neil Harl, Chuck Hassebrook, Hans Herren, Laura Jackson, Major General Michael Jeffery, Jenny Kao-Kniffen, Amanda Kimble-Evans, John Klironomos, Chad Kruger, Drake Larsen, Mark Liebig, Joe Logan, Gabrielle Ludwig, Charlie Maslin, Kathleen Masterson, David Miller, Amelia Moore, Jeff Moyer, Jeff Nelson, Bronwyn Nicholas, Dave Nomsen, Viveca Novak, Martin Nowak, Birju Pandya, Robert Parkhurst, Himadri Pakrasi, Yadu Pokhrel, John Reaganold, Debbie Reed, Rob Rex, Marlyn Richter, Steve Richter, Jo Robinson, Ashley Rood, Dan Rooney, Bob Rutherford, Michael Ryan, Mike Sands, Tracey Schohr, Sara Scherr, Tim Schwab, Whendee Silver, Johan Six, Mike Small, Bill Sproul, Ryan Stockwell, Fred Stokes, Robert Taylor, Richard Teague, Reyes Tirado, Luane Todd, Paul Towers, Peter Traverse, Diana Wall, Alan Wentz, Keith Wiebe, and David Zartmann. Many thanks also to Eric Strattman and the Knight Science Journalism Fellowship boot camp at MIT.

Years and years ago, a writer friend told me about a great editor she'd worked with named Alex Postman and made an introduction. Sadly, I never managed to come up with a good idea for that magazine or any of the others Alex worked for. I was thrilled when she turned up at Rodale, right around the time that I wrote the proposal for *The Soil Will Save Us*. My friend was right: Alex is the kind of editor that writers pine for, who in every instance made my manuscript smarter and cleaner and better and who was always cheerful, even when I was woefully late. (And many thanks to Marilyn Hauptly for putting up with the lateness and making nimble adjustments.)

Thanks also to Rodale's excellent cadre of fact-checkers, who saved me from error and mortification. Really, I wrote Pat Buchanan instead of James Buchanan?

And huge thanks to my wonderfully smart and capable agent, Kirsten Neuhaus. I love anyone who gets excited about the life in the soil and its ancient partnership with plants, and Kirsten caught it, right away. She was an inspiration in that she launched her business in the midst of the economic crisis, when so many people were giving up on publishing entirely. She not only believed in my book, but in the industry!

I am blessed (odd word for a kind-of atheist, but there you have it) with many friends, and lots of them are writers. They have enough of their wordy work to keep them busy, but a bunch of them took time, when asked, to read parts of the manuscript and tell me if it would interest and make sense to anyone who wasn't already a soil geek. So thanks again to Jill Adams, Barbara Benson, Kathleen Conroy, Bobbi Dempsey, Rachel Dickinson, Charlotte Huff, Mary Grimm, Susan Grimm, Marina Krakovsky, Gwen Moran, Mary Norris, Pam Oldham, Cynthia Ramnarace, Tricia Springstubb, and Anne Trubek. Thanks to Karen Long for wise counsel and encouragement before, during, and after the writing of this book. Finally, thanks to my first readers, Susan Lubell and my much-cherished daughter, Jamie Newell.

REFERENCES

CHAPTER 1

Lal R. "Carbon Emissions from Farm Operations." *Environment International* 30:981–990, 2004.

———. "Controlling Greenhouse Gases and Feeding the Globe through Soil Management." University Distinguished Lecture, Ohio State University, Columbus, February 17, 2000.

———. "Laws of Sustainable Soil Management." *Agronomy for Sustainable Development* 29:7–9, 2009.

———. "Managing Soils for Feeding a Global Population of 10 Billion." *Journal of the Science of Food and Agriculture* 86(14):2273–2284, 2006.

———. "Potential of Desertification Control to Sequester Carbon and Mitigate the Greenhouse Effect." *Climatic Change* 15:35–72, 2001.

———. "Ten Tenets of Sustainable Soil Management." *Journal of Soil and Water Conservation* 64(1):20A–21A, 2009.

Revelle R and Suess HE, "Carbon Dioxide Exchange Between Atmosphere and Ocean and the Question of an Increase in Atmospheric CO_2 During the Past Decades," *Tellus*, Vol. 9, Issue 1, 1957.

Weisenburger FP. *The Passing of the Frontier, 1825–1850*. The History of the State of Ohio, vol. 3. Columbus: Ohio State Archaeological and Historical Society, 1941.

CHAPTER 2

Blankenship RE, Raymond J, et al. "Evolution of Photosynthetic Antennas and Reaction Centers." *PS2001 Proceedings: 12th International Congress on Photosynthesis, PL13*. Melbourne: CSIRO, 2001.

Jones C. "Carbon That Counts" Presentation, New England and North West Landcare Adventure, Guyra, New South Wales, March 16–17, 2011.

———. "The Back Forty Down Under: Adapting Farming to Climate Variability." *Quivira Coalition Journal* 35:11–16, 2010.

Nowak MA, and Ohtsuki H. "Prevolutionary Dynamics and the Origin of Evolution." *Proceedings of the National Academy of Sciences of the United States of America* 105:14924–14927, 2008.

Sugden A, Stone R, et al., eds. "Soils: The Final Frontier." Special issue, *Science* 304(5677):1613–1637, 2004.

Taylor TN. "Fungal Associations in the Terrestrial Paleoecosystem." *Trends in Ecology and Evolution* 5(1):21–25, 1990.

Wall DH, Bardgett RD, and Kelly E. "Biodiversity in the Dark." *Nature Geoscience* 3:297–298, 2010.

CHAPTER 3

Hadley CJ. "The Wild Life of Allan Savory." *Range Magazine* Fall 1999, 44–47.

Ruddiman WF. "Millennia of Agricultural Resilience." The *Geographer* Autumn 2012, 8. http://www.rsgs.org/publications/TheGeographer-Autumn2012.pdf.

———. *Plows, Plagues, and Petroleum: How Humans Took Control of Climate.* Princeton, NJ: Princeton University Press, 2005.

Savory A, and Lambrechts J. "Holism: The Future of Range Science to Meet Global Challenges." *Grassroots* 12(3):28–47, 2012.

Savory A. *Holistic Management: A New Framework for Decision Making.* With Jody Butterfield. Washington, DC: Island Press, 1999.

Weber KT, and Gokhale BS. "Effect of Grazing on Soil-Water Content in Semiarid Rangelands of Southeast Idaho." *Journal of Arid Environments* 75(5):464–470, 2011.

Weber KT, and Horst S. "Desertification and Livestock Grazing: The Roles of Sedentarization, Mobility and Rest." *Pastoralism: Research, Policy and Practice* 1:19, 2011.

CHAPTER 4

Grogg P. "No-Till Farming Holds the Key to Food Security." Inter Press Service News Agency, February 20, 2013.

Pokhrel YN, Hanasaki N, et al. "Model Estimates of Sea-Level Change Due to Anthro-pogenic Impacts on Terrestrial Water Storage." *Nature Geoscience* 5:389–392, 2012.

Pollan M. *The Omnivore's Dilemma.* New York: Penguin Press, 2006.

Stockwell R, and Bitan E. *Future Friendly Farming: Seven Agricultural Practices to Sustain People and the Environment.* Reston, VA: National Wildlife Federation, 2011.

Stockwell R, and Bryant L. *Roadmap to Increased Cover Crop Adoption.* Reston, VA: National Wildlife Federation, 2012.

Sugden A, Stone R, et al., eds. "Soils: The Final Frontier." Special issue, *Science* 304(5677):1613–1637, 2004.

World Economic Forum. "What If the World's Soil Runs Out?" Time.com, December 14, 2012. http://world.time.com/2012/12/14/what-if-the-worlds-soil-runs-out.

CHAPTER 5

Chambers AS. "Encouraging Carbon Sequestration on Private Agricultural Lands in the United States." Extended Abstract #35, presented at Greenhouse Gas Strategies in a Changing Climate International Conference of the Air and Waste Management Association, San Francisco, November 16–17, 2011.

Coalition on Agricultural Greenhouse Gases. *Carbon and Agriculture: Getting Measurable Results*. Version 1. April 2010.

Foley J. "It's Time to Rethink America's Corn System." Ensia.com, March 5, 2013. http://ensia.com/voices/its-time-to-rethink-americas-corn-system.

Kragt ME, Pannell DJ, et al. "Assessing Costs of Soil Carbon Sequestration by Crop-Livestock Farmers in Western Australia." *Agricultural Systems* 112:27–37, 2012.

Laskawy T. "USDA Downplays Own Scientist's Research on Ill Effects of Monsanto Herbicide." Grist.org, April 21, 2010. http://grist.org/article/usda-downplays-own-scientists-research-on-danger-of-roundup.

National Agricultural Statistics Service, USDA. *2007 Census of Agriculture: United States Data*. Washington, DC: USDA, February 4, 2009.

———. "U.S. Corn Acreage Up for Fifth Straight Year." June 28, 2013. www.nass.usda.gov/Newsroom/printable/06_28_13.pdf.

"Watershed Payments Topped $8.17 Billion in 2011." EcosystemMarketplace.com, January 17, 2013.

Zwick S. "The $8 Billion Bargain: How Watershed Payments Save Cities, Support Farms and Combat Climate Change." Forbes.com, January 17, 2013.

CHAPTER 6

Bromfield L. *Malabar Farm*. New York: Harper, 1948.

———. *Pleasant Valley*. New York: Harper, 1945.

Callicrate M, and Stokes F. "From Berkeley to Boston: Coming Together around Freedom, Fairness and Food." Organization for Competitive Markets, March 16, 2013. www.competitivemarkets.com/from-berkeley-to-boston.

"FDA Sued for Concealing Records on Arsenic in Poultry Feed." EcoWatch.com, May 13, 2013. http://ecowatch.com/2013/fda-sued-concealing-records-arsenic-in-poultry-feed.

Food and Water Watch. *Public Research, Private Gain: Corporate Influence over University Agricultural Research*. Washington, DC: Food and Water Watch, April 2012.

Fuglie KO, Heisey PW, et al. *Research Investments and Market Structure in the Food Processing, Agricultural Input, and Biofuel Industries Worldwide*. Economic Research Report #130. Economic Research Service, US Department of Agriculture, December 2011.

Goldenberg S. "Secret Funding Helped Build Vast Network of Climate Denial Thinktanks." *Guardian*, February 14, 2013.

Greene W. "Guru of the Organic Food Cult." *New York Times*, June 6, 1971.

Harl N. "Commentary on Concentration and Anti-Competitive Practices in the Seed and Chemical Industry." Public hearing, Department of Justice Antitrust Division, Ankeny, Iowa, April 3, 2010.

Heffernan W, and Hendrickson M. *Consolidation in the Food and Agriculture System*. Report to the National Farmers Union, February 5, 1999. http://www.foodcircles.missouri.edu/whstudy.pdf.

Jackson L. "The Farm as Natural Habitat." Transcript of the 2005 Shivvers Memorial Lecture, Iowa State University, October 19, 2005. www.leopold.iastate.edu/sites/default/files/pubs-and-papers/2005-10-farm-natural-habitat.pdf.

———. "Who 'Designs' the Agricultural Landscape?" *Landscape Journal* 27(1):23–40, 2007.

Lowe P. "Public Research for Private Interests." Harvest Public Media, December 10, 2012. http://harvestpublicmedia.org/article/1531/public-research-private-interests-beef-industry/5.

Masterson K, "Round and Round We Go," Harvestpublicmedia.org, accessed September 20, 2011.

Mayer A. "Corn Checkoff Keeps Corn at the Fore." Harvest Public Media, April 9, 2013. http://harvestpublicmedia.org/content/corn-checkoff-keeps-corn-fore#.UlQovhbR2JU.

Pennsylvania Department of Environmental Protection. "J. I. Rodale and the Rodale Family: Celebrating 50 Years as Advocates for Sustainable Agriculture." Environmental Education: Pennsylvania's Environmental Heritage: Pennsylvania's Environmental Leaders: Jerome Rodale. n.d. http://www.portal.state.pa.us/portal/server.pt/community/dep_home/5968.

Pepper IL, Gerba CP, et al. "Soil: A Public Health Threat or Savior?" *Critical Reviews in Environmental Science and Technology* 39:416–432, 2009.

Philpott T. "Big Ag Won't Feed the World." MotherJones.com, June 15, 2011. www.motherjones.com/tom-philpott/2011/06/vilsack-usda-big-ag.

———. "Why the Government Should Pay Farmers to Plant Cover Crops." MotherJones.com, January 12, 2013.

Reganold JP, Jackson-Smith D, et al. "Transforming U.S. Agriculture." *Science* 332(6030):670–671, 2011.

Robinson C, and Latham J. "The Goodman Affair: Monsanto Targets the Heart of Science." Independent Science News, May 21, 2013. http://independentsciencenews.org/science-media/the-goodman-affair-monsanto-targets-the-heart-of-science.

Rodale Institute. *The Farming Systems Trial: Celebrating 30 Years*. Kutztown, PA: Rodale Institute, n.d.

Salatin J. *Folks, This Ain't Normal: A Farmer's Advice for Happier Hens, Healthier People, and a Better World*. New York: Center Street, 2011.

Salvador RJ. "Food Choices: Modernity and the Responsibility of Eaters." GreenFireTimes.com, December 31, 2012. http://greenfiretimes.com/2012/12/food-choices/#.UlQzuhbR2JU.

Samsel A, and Seneff S. "Glyphosate's Suppression of the Cytochrome P450 Enzymes and Amino Acid Biosynthesis by the Gut Microbiome: Pathways to Modern Diseases." *Entropy* 15(4):1416–1463, 2013.

Tegtmeier EM and Duffy MD, "External Costs of Agricultural Production in the United States," *International Journal of Agricultural Sustainability,* Vol. 1, No. 1, 2004.

CHAPTER 7

Clancy K. *Greener Pastures: How Grass-Fed Beef and Milk Contribute to Healthy Eating.* Cambridge, MA: Union of Concerned Scientists, March 2006.

DeLonge MS, Ryals R, and Silver WL. "A Lifecycle Model to Evaluate Carbon Sequestration Potential and Greenhouse Gas Dynamics of Managed Grasslands." *Ecosystems* 16:962–979, 2013.

Gurian-Sherman D. *Raising the Steaks: Global Warming and Pasture-Raised Beef Production in the United States.* Cambridge, MA: Union of Concerned Scientists, February 2011.

Marty JT. "Effects of Cattle Grazing on Diversity in Ephemeral Wetlands." *Conservation Biology* 19(5): 1626–1632, 2005.

Nelson D. "Common Ground: Ranchers, Environmentalists and Policymakers Unite to Protect Water Quality on California Rangeland." *UC Davis Magazine* 29(4), Summer 2012. http://ucdavismagazine.ucdavis.edu/issues/su12/common_ground.html.

Nomsen D. "Global Climate Change's Inevitable Impact on Hunters and Wildlife." PheasantsForever.org, April 17, 2009. [press release]

Robinson J. "Health Benefits of Grass-Fed Products." EatWild.com, n.d. www.eatwild.com/healthbenefits.htm.

Ryals R, and Silver WL. "Effects of Organic Matter Amendments on Net Primary Productivity and Greenhouse Gas Emissions in Annual Grasslands." *Ecological Applications* 23:46–59, 2013.

White C. "The Fifth Wave: Agrarianism and the Conservation Response in the American West." *Resilience* 37:38–54, January 2012.

———. *Revolution on the Range: The Rise of a New Ranch in the American West.* Washington, DC: Island Press, 2008.

CHAPTER 8

Koerth-Baker M. "Bloom Town: The Wild Life of American Cities." *New York Times Magazine,* December 2, 2012.

Presentation to Sandia National Laboratories by David C. Johnson, Institute for Sustainable Agricultural Research at New Mexico State University, August 14, 2012.

"Research Reveals Soil Carbon Capture Potential." *New Civil Engineer,* August 16, 2012.